AF360837

Cooperative Microactuator Systems

Cooperative Microactuator Systems

Editors

Manfred Kohl
Stefan Seelecke
Ulrike Wallrabe

MDPI • Basel • Beijing • Wuhan • Barcelona • Belgrade • Manchester • Tokyo • Cluj • Tianjin

Editors

Manfred Kohl
Karlsruhe Institute of
Technology
Karlsruhe, Germany

Stefan Seelecke
Saarland University
Saarbrücken, Germany

Ulrike Wallrabe
University of Freiburg
Freiburg, Germany

Editorial Office
MDPI
St. Alban-Anlage 66
4052 Basel, Switzerland

This is a reprint of articles from the Special Issue published online in the open access journal *Actuators* (ISSN 2076-0825) (available at: https://www.mdpi.com/journal/actuators/special_issues/microactuator_systems).

For citation purposes, cite each article independently as indicated on the article page online and as indicated below:

LastName, A.A.; LastName, B.B.; LastName, C.C. Article Title. *Journal Name* **Year**, *Volume Number*, Page Range.

ISBN 978-3-0365-8234-4 (Hbk)
ISBN 978-3-0365-8235-1 (PDF)

Cover image courtesy of Lisa Schmitt

Contents

About the Editors . . . . **vii**

Preface to "Cooperative Microactuator Systems" . . . . **ix**

Lisa Schmitt, Philip Schmitt and Martin Hoffmann
3-Bit Digital-to-Analog Converter with Mechanical Amplifier for Binary Encoded Large
Displacements
Reprinted from: *Actuators* **2021**, *10*, 182, doi:10.3390/act10080182 . . . . **1**

Almothana Albukhari and Ulrich Mescheder
Investigation of the Dynamics of a 2-DoF Actuation Unit Cell for a Cooperative Electrostatic
Actuation System
Reprinted from: *Actuators* **2021**, *10*, 276, doi:10.3390/act10100276 . . . . **17**

Lisa Schmitt and Martin Hoffmann
Large Stepwise Discrete Microsystem Displacements Based on Electrostatic Bending Plate
Actuation
Reprinted from: *Actuators* **2021**, *10*, 272, doi:10.3390/act10100272 . . . . **41**

Anna Christina Thewes, Philip Schmitt, Philipp Löhler and Martin Hoffmann
Design and Characterization of an Electrostatic Constant-Force Actuator Based on a Non-Linear
Spring System
Reprinted from: *Actuators* **2021**, *10*, 192, doi:10.3390/act10080192 . . . . **55**

Kirill Poletkin
On the Static Pull-In of Tilting Actuation in Electromagnetically Levitating Hybrid
Micro-Actuator: Theory and Experiment
Reprinted from: *Actuators* **2021**, *10*, 256, doi:10.3390/act10100256 . . . . **69**

Lena Seigner, Georgino Kaleng Tshikwand, Frank Wendler and Manfred Kohl
Bi-Directional Origami-Inspired SMA Folding Microactuator
Reprinted from: *Actuators* **2021**, *10*, 181, doi:10.3390/act10080181 . . . . **83**

**Gowtham Arivanandhan, Zixiong Li, Sabrina M. Curtis, Lisa Hanke, Eckhard Quandt and
Manfred Kohl**
Power Optimization of TiNiHf/Si Shape Memory Microactuators
Reprinted from: *Actuators* **2023**, *12*, 82, doi:10.3390/act12020082 . . . . **95**

Michael Olbrich, Arwed Schütz, Tamara Bechtold and Christoph Ament
Design and Optimal Control of a Multistable, Cooperative Microactuator
Reprinted from: *Actuators* **2021**, *10*, 183, doi:10.3390/act10080183 . . . . **105**

**Sipontina Croce, Julian Neu, Jonas Hubertus, Stefan Seelecke, Guenter Schultes and
Gianluca Rizzello**
Model-Based Design Optimization of Soft Polymeric Domes Used as Nonlinear Biasing Systems
for Dielectric Elastomer Actuators
Reprinted from: *Actuators* **2021**, *10*, 209, doi:10.3390/act10090209 . . . . **125**

Arwed Schütz, Sönke Maeter and Tamara Bechtold
System-Level Modelling and Simulation of a Multiphysical Kick and Catch Actuator System
Reprinted from: *Actuators* **2021**, *10*, 279, doi:10.3390/act10110279 . . . . **151**

About the Editors

Manfred Kohl

Manfred Kohl received his Ph.D. in Physics from the University of Stuttgart, Germany, in 1989. Later on, he worked as an IBM postdoctoral fellow at the T.J. Watson Research Center in Yorktown Heights, USA, and subsequently joined the Karlsruhe Institute of Technology (KIT), Germany, in 1992. He is currently a Professor in the Faculty of Mechanical Engineering and Head of the Department of Smart Materials and Devices at the Institute of Microstructure Technology of KIT. He is the spokesperson of the Priority Program KOMMMA (Cooperative Multistage Multistable Microactuator Systems) of the German Science Foundation. His current research focuses on ferroelastic and ferromagnetic shape memory alloys, multimaterial micro- and nanotechnologies, as well as corresponding smart devices.

Stefan Seelecke

Stefan Seelecke received a Ph.D. in Engineering Science from Technical University Berlin, Berlin, Germany, in 1995. After his habilitation in 1999, he joined the Department of Mechanical and Aerospace Engineering, North Carolina State University, Raleigh, NC, USA, in 2001. He is currently a Full Professor of Systems Engineering and Materials Science and Engineering at Saarland University, Saarbrücken, Germany, where he directs the Intelligent Material Systems Lab. His research interests include the development of smart-material-based actuator and sensor systems, in particular, (magnetic) shape memory alloys, piezoelectrics, and electroactive polymers.

Ulrike Wallrabe

Ulrike Wallrabe studied physics at Karlsruhe University (today, KIT), Germany. In 1992, she received her PhD degree in Mechanical Engineering of Microturbines and Micromotors. From 1989 to 2003, she was affiliated with the Institute of Microstructure Technology at Forschungszentrum Karlsruhe (today, KIT) working on microactuators and optical MEMS. Since 2003, she has held a Professorship for Microactuators in the Department of Microsystems Engineering, IMTEK, at the University of Freiburg, Germany. Her research focuses lie in adaptive optics, using piezo actuators to tune elastic lenses and mirrors, and in wireless energy transfer via electromagnetic waves or ultrasound. From 2012 to 2019, she was a member of the Cluster of Excellence BrainLinks-BrainTools, and since 2022 she has been a co-spokesperson in the Collaborative Research Centre (SFB-1537) ECOSENSE.

Preface to "Cooperative Microactuator Systems"

Ongoing miniaturization and the increase in demands for the functionality of microsystems generate an urgent need for innovative approaches in the control of, e.g., mechanics, optics, or fluidics on a miniature scale. An important prerequisite is the smart coupling of distributed microactuators to a cooperative synergetic actuation system. This opens up the potential to generate new functionalities and thereby to fulfill complex tasks comprising combinations of force, displacement, and dynamics that have not been possible until now. In the meantime, many different microactuators have been developed and are in use. Yet, their systematic coupling to cooperative multistage or, for instance, multistable microactuator systems still needs to be explored.

This Special Issue collects the emerging research activities in the field of cooperative microactuator systems, which are expected to generate new synergies, e.g., through parallelization, cascading, and multistability, as well as through inherent sensing. The presented contributions have come from the Priority Program KOMMMA (Cooperative Multistage Multistable Microactuator Systems) of the German Science Foundation. This interdisciplinary program addresses the various challenges of cooperative microactuator systems by bringing together research groups in the different research areas of microactuators, microsystems, material science, system simulation, and control and systems engineering. The Special Issue covers a broad range of microactuator systems based on electrostatic, shape memory, magnetic levitation, and dielectric elastomer principles. Furthermore, new methods for design and optimal control are introduced and new concepts for the modeling and simulation of the multiphysical properties of microactuator systems are presented.

Manfred Kohl, Stefan Seelecke, and Ulrike Wallrabe

Editors

Article

3-Bit Digital-to-Analog Converter with Mechanical Amplifier for Binary Encoded Large Displacements

Lisa Schmitt *, Philip Schmitt and Martin Hoffmann *

Faculty of Electrical Engineering and Information Technology, Ruhr-University Bochum (RUB), 44801 Bochum, Germany; philip.schmitt@rub.de
* Correspondence: lisa.schmitt-mst@rub.de (L.S.); martin.hoffmann-mst@rub.de (M.H.)

Abstract: We present the design, fabrication, and characterization of a MEMS-based 3-bit Digital-to-Analog Converter (DAC) that allows the generation of large displacements. The DAC consists of electrostatic bending-plate actuators that are connected to a mechanical amplifier (mechAMP), enabling the amplification of the DAC output displacement. Based on a parallel binary-encoded voltage signal, the output displacement of the system can be controlled in an arbitrary order. Considering the system design, we present a simplified analytic model, which was confirmed by FE simulation results. The fabricated systems showed a total stroke of approx. 149.5 ± 0.3 µm and a linear stepwise displacement of 3 bit correlated to $2^3 \hat{=}$ eight defined positions at a control voltage of 60 V. The minimum switching time between two input binary states is 0.1 ms. We present the experimental characterization of the system and the DAC and derive the influence of the mechAMP on the functionality of the DAC. Furthermore, the resonant behavior and the switching speed of the system are analyzed. By changing the electrode activation sequence, 27 defined positions are achieved upgrading the 3-bit systems into a 3-tri-state (3^3) system.

Keywords: digital-to-analog converter; DAC; large displacements; electrostatic bending-plate actuator; mechanical amplifier; mechAMP; 3-tri-state system; dicing-free SOI-technology; MEMS

Citation: Schmitt, L.; Schmitt, P.; Hoffmann, M. 3-Bit Digital-to-Analog Converter with Mechanical Amplifier for Binary Encoded Large Displacements. *Actuators* **2021**, *10*, 182. https://doi.org/10.3390/act10080182

Academic Editors: Manfred Kohl, Stefan Seelecke and Ulrike Wallrabe

Received: 30 June 2021
Accepted: 31 July 2021
Published: 4 August 2021

Publisher's Note: MDPI stays neutral with regard to jurisdictional claims in published maps and institutional affiliations.

1. Introduction

The presented cooperative multistage multi-stable actuator system with large stroke was motivated by terahertz (THz) applications where a free-space true-time delay is required. Usually, the true-time delay is subdivided in binary fractions of the wavelength. Mostly, MEMS-based reflect arrays for beam steering applications are driven by out-of-plane actuation [1], but this article focusses on an in-plane single-reflector system with a large displacement. The single in-plane chips are intended to be stacked together after fabrication to realize the mechanical beam steering in the THz range [2,3].

A digital actuator system would be a perfect solution to realize the in-plane discretized large throw. In order to avoid complex driving electronics, a micro electromechanical digital-to-analog converter (M-DAC) based on cascaded actuators directly addressed by the digital signal is presented. A mechanical DAC converts an electrical binary code into a proportional output displacement [4]. These DACs consist of n electrostatic actuators reaching a finite number of 2^n positions inherent to the actuators. By adding more actuators, the positioning performance tends to approach the positioning performance of continuous systems [4]. The M-DACs allow accurate, stable and repeatable displacements without extensive feedback control circuits or strict control requests [4]. In contrast to actuators driven by analog voltages, such systems do not need a careful position control including sensing.

Apart from beam steering applications, M-DACs can be used for open-loop operation [5] with high-position resolution and for optical surface profiling [6]. Yeh et al. [7] proposed a thermal actuation for a 3-bit maximum stroke of 5.8 µm at 5 V control voltage (Table 1). Liu et al. [8] presented a 4-bit weighted stiffness micro-electromechanical DAC with a maximum stroke of 3.75 μm and a control voltage of 5 V. Pandiyan et al. [9] suggested two different concepts based on a bent beam electrothermal compliant actuator as a single mechanical bit reaching maximum strokes of 44 μm and 45 μm at 5 V. As thermal actuators induce heating of the entire microsystem, these actuators are not practicable for applications requiring high precision positioning and quick response times. The drawback of electrostatic actuators is the comparable high voltage required for a large displacement. However, the works of Toshiyoshi et al. [10] and Sarajic et al. [4,11] showed that a strategic optimization of the system setup can lower the control voltage and increase the total stroke. By using comb-drive actuators, Toshiyoshi et al. and Sarajic et al. reached binary encoded displacements of 5.8 μm at 150 V and 8.57 μm at 45 V, respectively.

Table 1. Comparison of digital-to-analog converters from the literature.

Ref.	Actuation	No. of Bits	Maximum Stroke	Voltage
[7]	thermal	3	5.8 μm	5 V
[8]	electrothermal compliant	4	3.75 μm	5 V
[9]	electrothermal compliant	4	44 μm/45 μm	5 V/5 V
[10]	electrostatic (comb-drive)	4	5.8 μm	150 V
[4,11]	electrostatic (comb-drive)	12	8.57 μm	45 V
this work	electrostatic (bending-plate)	3/3-tri-states	DAC + mechAMP: 149.5 μm only DAC: 21.4 μm	60 V/75 V

In this paper, we present an in-plane 3-bit DAC consisting of cascaded flexible electrostatic bending-plate actuators for large displacement applications at low voltage. We report the combination of a DAC with a mechanical amplifier (mechAMP) [12] to generate large strokes of up to 149.5 μm at 60 V, including the control of $2^3 \triangleq 8$ discrete positions. By changing the electrode activation sequence, we can approach three states with every bit, thus enlarging the system application field to a 3-tri-state system driving $3^3 \triangleq 27$ discrete defined positions.

In Section 2 the single system components and the system concept are presented. Based on this, we derive the analytic transfer function and design the system supported by FE simulation. Section 3 shows the SOI fabrication process. Section 4 presents the experimental setup and the characterization results. We characterize the function of the DAC, of the mechAMP and of the combined system; the switching speed; the resonant behavior and the expansion of the 3-bit system to a 3-tri-state system. The main achievements of this work are summarized in Section 5.

2. Concept

2.1. System Components

The displacement of the system is generated by combining high-throw electrostatic bending-plate actuators [2,13]. These actuators are illustrated in Figure 1a and consist of long, nonrigid cantilevers and electrodes with low stiffness.

The cantilever of the fixed actuator (f3) is clamped. The cantilever of the moving actuator (m2) is directly linked to a linear rotor guided and fixed by a flexible spring. The electrodes (m1, f1) face each other at a distance $x_{electrode}$. When applying a voltage between the electrodes, the cantilevers and electrodes bend down as soon as the electric force attracting the electrodes exceeds the mechanical force of the cantilevers, electrodes and the spring. At low voltage, the tips of the electrodes (m1, f1) begin to approach. With increasing voltage, the pull-in is completed, as illustrated in Figure 1b.

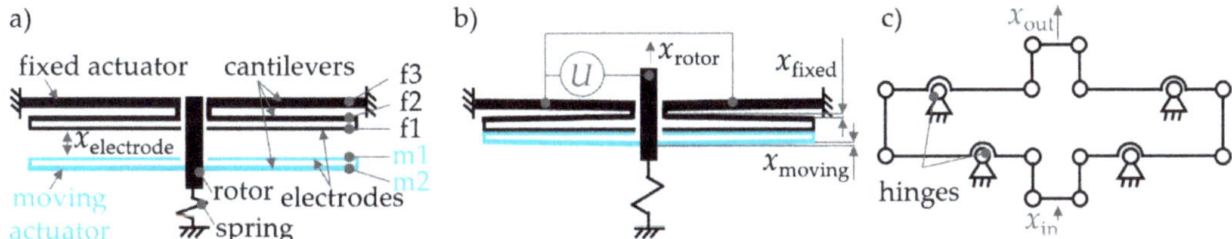

Figure 1. (**a**) Setup of a high-throw electrostatic bending-plate actuator. (**b**) Bending-plate actuator traveling the distance x_{rotor} after the application of a voltage. (**c**) Setup of the mechAMP converting a small input displacement x_{in} into a large output displacement x_{out}. (drawings not to scale).

The distance covered by the linear rotor x_{rotor} corresponds to the electrode gap $x_{\text{electrode}}$ before pull-in, deducting the deflection of the fixed and the moving actuators x_{moving} and x_{fixed} (Figure 1b), which yields:

$$x_{\text{rotor}} = x_{\text{electrode}} - (x_{\text{fixed}} + x_{\text{moving}}) \tag{1}$$

The displacement of the electrostatic actuators is amplified by a passive micromechanical amplifier (mechAMP) [12], as shown in Figure 1c. It consists of cascaded levers connected by flexure hinges. The amplification ratio A_{mechAMP} is designed by adjusting the hinges to convert a small input displacement x_{in} into a large output displacement x_{out}. The input stiffness of the mechAMP is defined by the length of the beams and by the thickness and the position of the flexure hinges.

2.2. System Function

The system (Figure 2a) consists of two components: a mechAMP and a 3-bit DAC driven by multiple electrostatic actuators. The binary encoded discrete displacement (small stroke but high force) generated by the DAC x_{DAC} is amplified by the mechAMP, resulting in a large and still binary-encoded discrete total system displacement x_{system}.

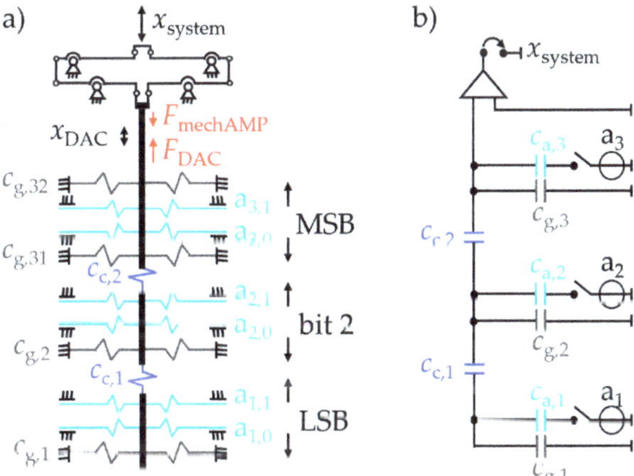

Figure 2. (**a**) Structure of the system: the binary encoded discrete displacement of the DAC x_{DAC} is amplified by the mechAMP, resulting in a large and still binary-encoded discrete system displacement x_{system}, (**b**) corresponding electrical network model. (drawings not to scale).

The 3-bit DAC (Figure 2a) consists of multiple electrostatic actuators a_i with a stiffness $c_{a,i}$. Each bit has actuators $a_{i,0}$ that can travel downwards (0-direction) and actuators $a_{i,1}$ that can travel upwards (1-direction). The electrode gaps $x_{\text{electrode},i}$ are identical for each actuator. The bits are connected by springs $c_{c,i}$. Sinusoidal guiding springs $c_{g,i}$ connect the system to the substrate. Bit 3, the most significant bit (MSB), is connected to the mechAMP. To increase the stability of the mechAMP and the MSB, the MSB is connected to the substrate by two guiding springs, $c_{g,32}$ and $c_{g,31}$, whereas the other two bits feature only one guiding spring, $c_{g,1}$ and $c_{g,2}$, respectively.

Figure 3 exemplarily shows the different states of the DAC. Each bit is either shifted upwards or downwards. When all downwards-actuators ($a_{i,0}$) are displaced, the system is in position 000 (Figure 3a). The system obtains the next logical position (001) when the upward actuators of the least significant bit (LSB, bit 1) $a_{1,1}$ are activated and the downwards actuators $a_{1,0}$ are deactivated at the same time (Figure 3b). When all upwards actuators ($a_{i,1}$) are displaced, the system is in position 111 (Figure 3c).

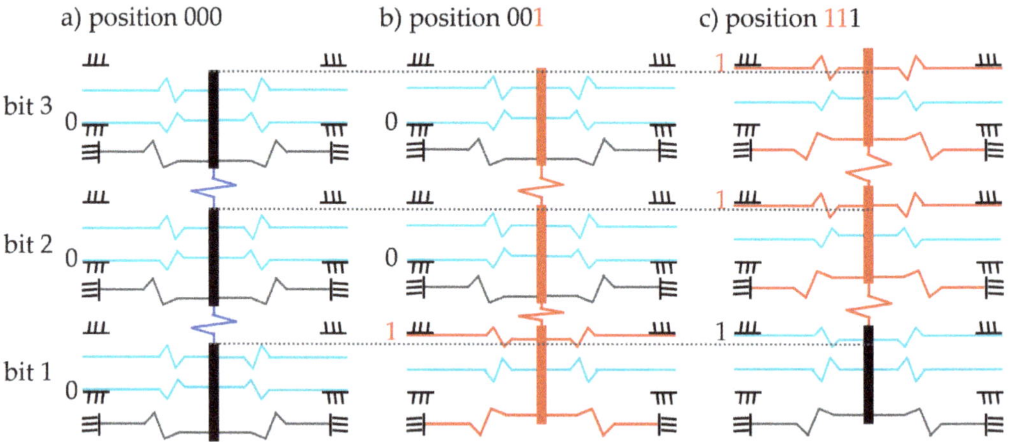

Figure 3. DAC in (**a**) the 000 position by activating the $a_{i,0}$ actuators; (**b**) the 001 position by activating the $a_{1,1}$, $a_{2,0}$ and $a_{3,0}$ actuators and (**c**) the 111 position by activating the $a_{i,1}$ actuators. (drawings not to scale).

In an electrical network model (Figure 2b), the actuators are modeled as potential sources, and the mechanical springs are associated with the capacitors. The network model allows the derivation of the analytic transfer function (2) for the DAC [2]:

$$
\begin{pmatrix}
c_a + c_g + c_c & -c_c & 0 \\
-c_c & c_a + c_g + 2c_c & -c_c \\
0 & -c_c & c_a + c_g + c_c
\end{pmatrix}
\cdot
\begin{pmatrix}
x_{\text{rotor},1} \\
x_{\text{rotor},2} \\
x_{\text{rotor},3}
\end{pmatrix}
=
\begin{pmatrix}
F_{a,1} \\
F_{a,2} \\
F_{a,3}
\end{pmatrix}
\tag{2}
$$

Equation (2) is valid as long as each bit features a displacement either corresponding to a 0- or to 1-direction ($x_{a,1,0} \neq x_{a,1,1}$, $x_{a,2,0} \neq x_{a,2,1}$ and $x_{a,3,0} \neq x_{a,3,1}$), assuming each actuator has the same stiffness ($c_{a,1} = c_{a,2} = c_{a,3}$) and travels the same distance ($x_{\text{rotor},1} = x_{\text{rotor},2} = x_{\text{rotor},3}$). All connecting springs ($c_{c,1} = c_{c,2} = c_{c,3}$), as well as all guiding springs ($c_{g,1} = c_{g,2} = c_{g,3}$), have the same stiffness [2].

As illustrated in Figure 2a, the mechanical force of the DAC F_{DAC} must overcome the stiffness of the mechAMP F_{mechAMP}. For $F_{\text{DAC}} \gg F_{\text{mechAMP}}$, the system displacement yields:

$$
x_{\text{system}} = x_{\text{DAC}} \cdot A_{\text{mechAMP}}
\tag{3}
$$

2.3. System Modeling and Simulation

The function of the system and its components is verified by solid-state mechanic and electrostatic finite-element (FE) simulations carried out in COMSOL Multiphysics. The optimization is focused on a maximum stroke and a linear stepwise displacement.

A single electrostatic actuator is analyzed using 2D electrostatic stationery and time-dependent FE simulations based on the setup shown in Figure 4. The flexible electrodes are clamped at each end. The voltage between the electrodes is systematically increased. During simulation, a moving mesh is used. Spacers defined as contact pairs allow a steady simulation. These semicircular-formed spacers are adopted for the fabrication of the electrostatic actuators, as they minimize the contact area of the electrodes and, therefore, also the risk of a short circuit during experimental characterization. The simulation of the single actuator shows a pull-in voltage of the tips between 20 V and 30 V.

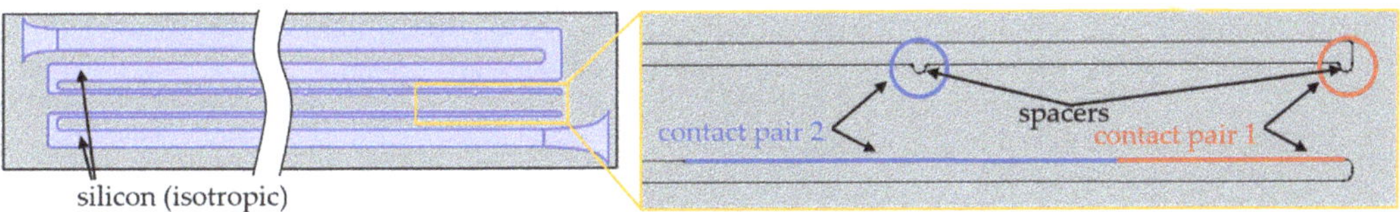

Figure 4. Simulation setup of a single actuator by COMSOL Multiphysics: a voltage is systematically increased between the electrodes, and the spacers defined as contact pairs allow a steady simulation.

As explained in Section 2.1, the cantilevers of the bending-plate actuators bend when a sufficiently high voltage is applied. To achieve a high displacement of the rotor x_{rotor} (Equation (1)), the stiffness of the fixed actuator must exceed the stiffness of the moving system. Therefore, the cantilever stiffness has to be sufficiently high to overcome the stiffness of the guiding and connecting springs. However, a higher stiffness of the cantilevers results in a higher pull-in voltage. To optimally balance the opposing requirements for low pull-in voltage and large displacement, multiple actuators connected in a parallel configuration are used. In this arrangement, the single actuator features a low pull-in voltage due to its low stiffness, but the total stiffness is increased by the connection of multiple actuators. Figure 5 shows such an actuator consisting of twelve single identical high-throw bending-plate actuators for each direction. The electrodes of the actuators have a thickness $t_{\text{a,m1/f1}} = 5$ µm and length $L_{\text{a,m1/f1}} = 2210$ µm. The lengths of the cantilevers amount to $L_{\text{a,f3}} = 2235$ µm and $L_{\text{a,m2}} = 2265$ µm, the thickness is $t_{\text{a,f3/m2}} = 25$ µm and $t_{\text{a,f2}} = 20$ µm.

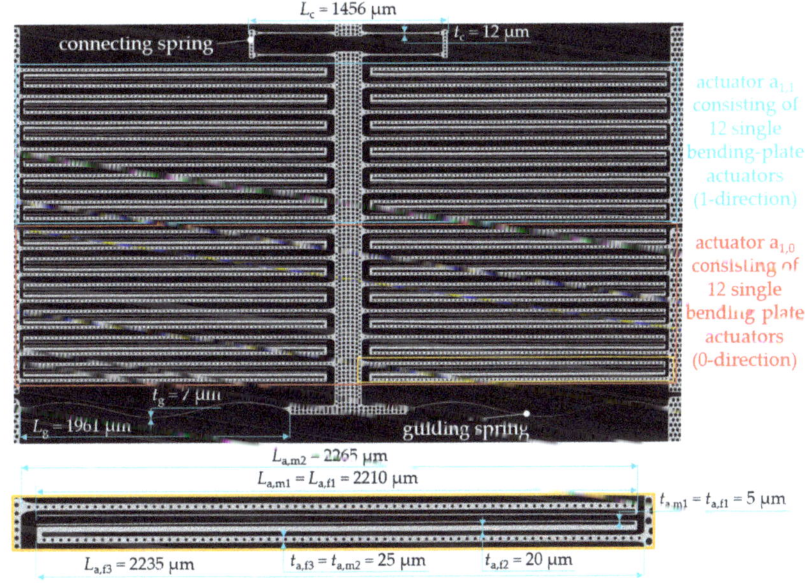

Figure 5. Bit 1 fixed by guiding springs and connected to bit 2 by a connecting spring. Each actuator, $a_{1,0}$ and $a_{1,1}$, consists of 12 single bending-plate actuators (stiching microscope image).

The guiding springs connect the DAC to the substrate and prevent an in-plane rotation. The sinusoidal guiding springs (Figure 5) are well-suited here, as they feature a very low stiffness in the deflection direction, lowering the pull-in voltage and, therefore, increasing the displacement of the actuators [14]. However, in off-axis directions, these springs show a high stiffness ensuring a purely translational guiding of the system [14]. The guiding springs feature the thickness t_g = 7 µm and the length L_g = 1.961 µm. The connecting springs (Figure 5) combine the single bits and feature a high stiffness to guarantee the transmission of the actuator displacement to the system output. The minimum thickness of the connecting springs t_c amounts to 12 µm, and the length L_c is 1456 µm.

For the mechanical amplification of the DAC output displacement, we use the two different concepts of displacement amplification presented in Figure 6. The function principle of mechAMP+ shown in Figure 6a is identical to the version presented in Figure 1c. The mechAMP+ features four hinges with a minimum thickness t_{hinge} = 3 µm and directly converts the small input displacement x_{in} in a large output displacement x_{out}. Instead of hinges, the mechAMP− in Figure 6b is guided by a triangular spring [14] at the output. The spring features 26 beams with an inclination angle α = 60° and has a thickness t_{spring} of 5 µm and a length L_{spring} of 2952 µm. This spring provides additional stability for the connection of the mechAMP to the substrate. Therefore, the mechAMP− does not only convert the small input displacement x_{in} into a large output displacement x_{out}, it also inverts the input signal. As illustrated in Figure 6b, a positive input displacement results in a negative output displacement and vice versa.

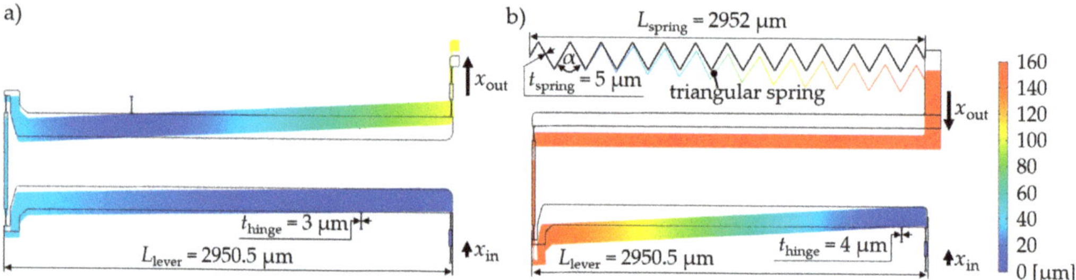

Figure 6. MechAMPs displaced the distance x_{in} = 10 µm by the COMSOL Multiphysics simulation. Due to the symmetric setup, only the left hand side of the mechAMPs is shown. (**a**) MechAMP+ amplifies the input displacement x_{in} in a large output displacement x_{out}, and (**b**) MechAMP− amplifies and inverts the input displacement x_{in} in a large and inverse output displacement x_{out}.

Based on the presented considerations, we model two systems consisting of different mechAMPs and DACs, shown in Table 2. The modeling results refer to an electrode displacement of 10 µm. The sinusoidal guiding springs feature a stiffness of c_g = 0.7 N/m, the connecting springs of c_c = 128 N/m and the actuators of c_a = 92 N/m. DAC 1 features an additional spring connecting the LSB to the substrate (Figure 7i).

The modeled systems show maximum strokes $x_{system,max}$ of 143.7 µm and 167.1 µm, respectively. The step sizes amount to 20.5 ± 6.7 µm and 23.9 ± 3.1 µm, respectively. DAC 1 and DAC 2 achieve maximum strokes of $x_{DAC1,max}$ = 17.8 µm and $x_{DAC2,max}$ = 19.8 µm, respectively. MechAMP+ features an amplification ratio of 10.3 with an input stiffness of 44.7 µN. MechAMP+ lowers the maximum stroke of DAC 1 to 14.0 µm. MechAMP− has a higher amplification ratio of 15.1 but, also, a higher input stiffness of 127.4 µN, lowering the maximum stroke of DAC 2 to 11.1 µm. The size of system 1 amounts to 10,500 × 5980 µm², and the chip size amounts to 15,165 × 10,930 µm² (Table 2 and Figure 7i).

Table 2. Simulated DACs, mechAMPs and systems at a displacement of $x_{\text{electrode}} = 10$ µm.

		System 1 (Figure 7i)	**System 2**
	version	*DAC 1*	*DAC 2*
DAC	$x_{\text{DAC,max}}$	17.8 µm	19.8 µm
	step size	2.5 ± 0.8 µm	2.8 ± 0.4 µm
	version	*mechAMP+*	*mechAMP–*
mechAMP	amplification ratio	10.3	15.1
	input stiffness	44.7 µN	127.4 µN
DAC combined to	$x_{\text{DAC,max}}$	14.0 µm	11.1 µm
mechAMP	step size	2.0 ± 0.7 µm	1.6 ± 0.2 µm
	$x_{\text{system,max}}$	143.7 µm	167.1 µm
system	step size	20.5 ± 6.7 µm	23.9 ± 3.1 µm
	system size	10,500 × 5980 µm^2	11,080 × 6144 µm^2
	chip size	15,165 × 10,930 µm^2	15,707 × 10,930 µm^2

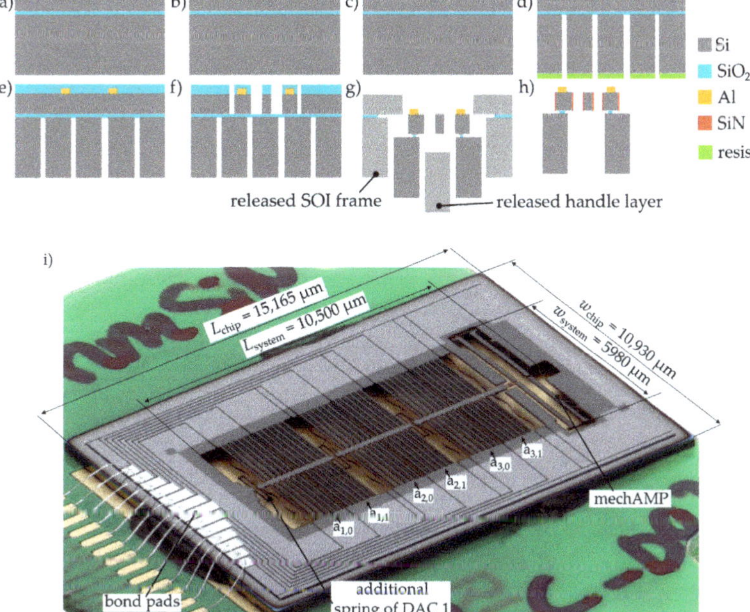

Figure 7. Fabrication process (drawings not to scale): (**a**) SOI Wafer, (**b**) deposition of the aluminum layer, (**c**) etching of the bond pads, (**d**) deep etching of the handle layer, (**e**) PECVD of 400-nm SiO₂, (**f**) deep etching of the device layer, (**g**) HF vapor etching, (**h**) deposition of 400-nm SiN and (**i**) fabricated chip fixed and wire-bonded to a circuit board (stacked device photo).

3. Fabrication

The microsystems are fabricated on (100)-oriented Silicon-on-Insulator (SOI) wafers (Figure 7a) with a 300-µm-thick handle layer and a 50-µm-device layer. To minimize the risk of a pull-in between the electrostatic actuators and the handle layer and to prevent a parasitic out-of-plane motion during electronic activation, the chips are fabricated by a dicing-free process [15,16]. This process separates the chips from each other, making the mechanical sawing process obsolete and, also, allows the removal of the handle layer from the backside of the electrostatic actuators during hydrofluoric (HF) vapor etch release.

First, a 100-nm aluminum layer is deposited on the device layer by evaporation (Figure 7b), and the bond pads are patterned by wet chemical etching (Figure 7c). Then, the resist (AZ MIR 701 29 cp) is spin-coated with 1000 rpm onto the handle layer as a mask during deep etching. For deep etching, an alternating SF_6/C_4F_8-based deep reactive ion etching process (DRIE) is used (Figure 7d). After O_2 plasma, a 400-nm SiO_2 layer is deposited by a plasma-enhanced chemical vapor deposition (PECVD) process (Figure 7e). The SiO_2 layer is used as a hard mask for the patterning of the structures into the device layer. Then, the structures are patterned and deep-etched (Figure 7f). HF vapor etching releases the moveable structures (Figure 7g). Then, the electrodes are isolated by depositing 400-nm silicon nitride (SiN) with a low-stress PECVD process (Figure 7h). During this process, the chips are flipped to reduce the spacing between the electrodes and the carrying wafer, which also prevents a coating of the bond pads with isolating SiN. In the final step, the chips are fixed to a circuit board and wire-bonded (Figure 7i).

4. System Characterization Results

4.1. Experiment and Characterization Setup

The systems are characterized using a highspeed camera system (Keyence VW-600C) and a voltage source (EA- Electro-Automatic PS-3200-02 C). The characterization setup is presented in Figure 8.

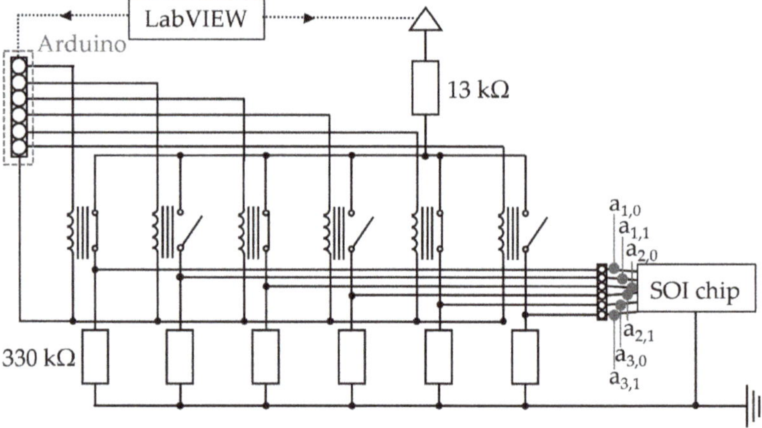

Figure 8. A LabVIEW program controls the voltage source, with an Arduino activating the relays in an arbitrary order and as a function of time. The relays are connected to the actuators of the SOI chip. Here, the relays are exemplarily activated to drive the 000 position.

The LabVIEW program controls the voltage source and an Arduino that drives the single bits of the microsystem by activating and deactivating the relays on a circuit board in an arbitrary order and as a function of time. The relays are connected to the actuators on the SOI chip. Pull-down resistors of 330 kΩ are connected between the electrostatic actuators and the ground potential to enable a quick discharge of the electrodes. The 13-kΩ resistor limits the current flow to prevent an eventual breakthrough of the SiN isolation at the electrodes. For the actuation of the electrostatic actuators, a DC voltage is used.

Based on the videos recorded by the highspeed camera during the activation of the DAC, the displacement of the system and its components is analyzed. Due to the very high aspect ratio (length vs. displacement) and due to the symmetric setup of the system, the videos show either the middle, the left or the right side of the system. To analyze the videos, the software *Tracker* allows marking of the strategic tracking points.

4.2. Characterization of the System

4.2.1. System Function Testing

Figure 9 illustrates the operation of the system. Depending on the activated actuators, the mechAMP and, with it, the output of the system is displaced from maximum (111 position) to the minimum (000 position) displacement.

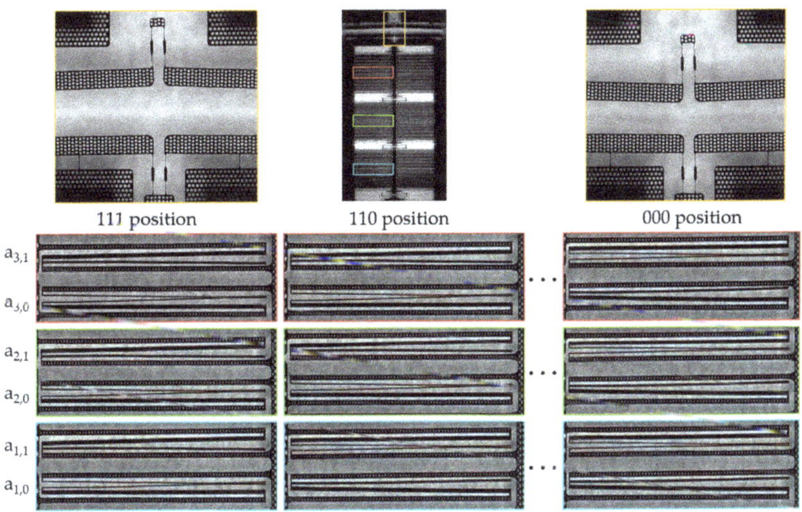

Figure 9. (**Top**) Middle: system 1, left: mechAMP+ in the 111 position, right: mechAMP+ in the 000 position. (**Bottom**) Left: actuators in the 111 position, middle: actuators in the 110 position, right: actuators in the 000 position. Due to the high aspect ratio, the length of the actuators is compressed by 50%.

4.2.2. Characterization of the DAC

In a first characterization step, the pull-in voltage of the single bits in relation to the driven position is analyzed in Figure 10. Each position is approached in logical order from the 111 to 000 position. Starting with an initial voltage of 30 V, the voltage is increased in 5 V steps, until a pull-in of all actuators of each single bit is achieved.

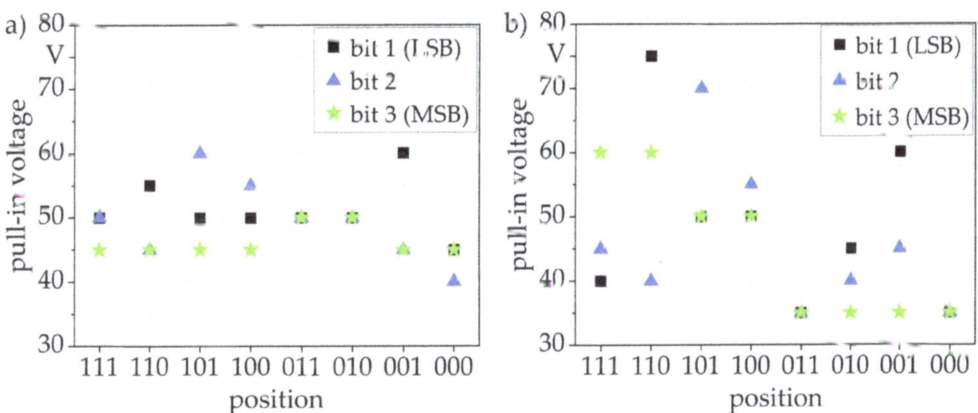

Figure 10. Pull-in voltage of the single bits is dependent on the driven position for (**a**) system 1 and (**b**) system 2. During the experiment, the voltage increased in 5 V-steps, until a pull-in of all the actuators of each single bit was achieved.

Bit 2 of system 1 (Figure 10a) shows the lowest pull-in voltage of 40 V at the 000 position. As bit 1 and bit 3 pull in at 45 V, the pull-in voltage at the 000 position amounts to 45 V. The 001 and 101 positions have the highest pull-in voltage of 60 V. Therefore, a control voltage of 60 V is chosen for system 1. Consequently, the control voltage of system 2 (Figure 10b) amounts to 75 V. The high input stiffness of mechAMP– can be considered as the cause for the higher control voltage of system 2.

At first glance, the derived control voltage of the system appears very high. In order to classify this result, a linear relationship between the displacement and voltage is assumed. System 1 reaches a total displacement of 150 µm at 60 V, which corresponds to 2.5 µm per 1 V. The electrostatic comb-drive DACs [4,10,11] presented in Table 1 reached a displacement of approx. 0.04 µm and 0.2 µm per 1 V, respectively. The electrostatic parallel-plate actuator presented in reference [13] reached a displacement of 65 µm at 30 V, which corresponds to 2.2 µm per 1 V. Therefore, in relation to the achieved displacement, the voltage of the presented system is rather low.

During the experimental characterization of the DAC, stiction issues appeared at low voltages. Thus, at the selected control voltage, no stiction issues were observed in the further experiments. Additionally, when varying the switching speed in the range from 2 s to 0.1 milliseconds, no stiction issues were observed.

Figure 10 shows that the pull-in voltage of the electrostatic actuators depends on the system, the bit and on the driven position. For system 1 (Figure 10a), the MSB usually exhibited a lower pull-in voltage than the LSB and bit 2. The stiffness of the additional connecting spring (Figure 7i) that the LSB was connected to was higher than the input stiffness of the mechAMP+ that the MSB was connected to. Therefore, the MSB had to overcome a lower mechanical stiffness than the LSB and had a lower pull-in voltage. Apart from positions 111 and 000, bit 2 always displaces in the opposite direction either of the MSB or of the LSB, which increased the pull-in the voltage of bit 2. At the 000 and 111 positions, the actuators tended to show a low pull-in voltage, so that the displacement of all actuators in the same direction seemed to support a low pull-in voltage.

The displacement of the actuators depending on the driven position was analyzed experimentally, as presented in Figures 11 and 12

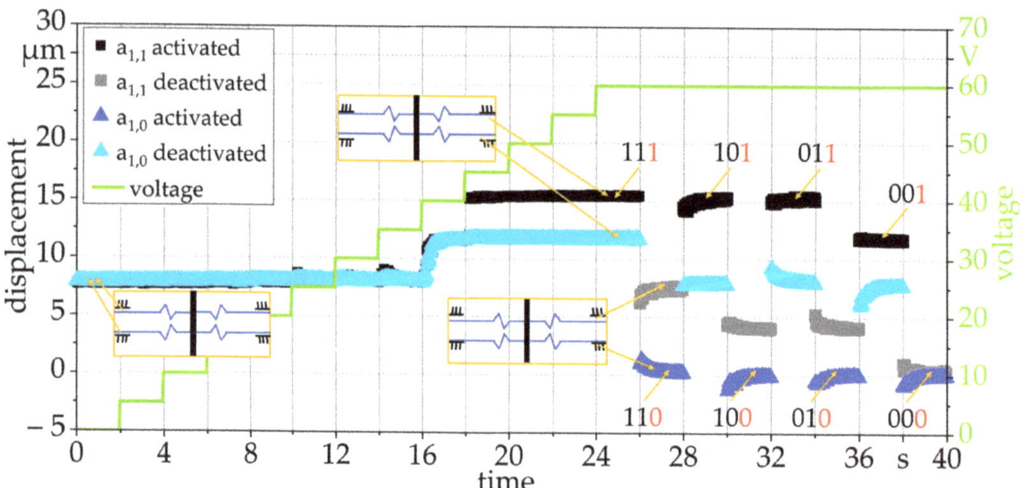

Figure 11. Displacement of the actuators $a_{1,1}$ and $a_{1,0}$ of the LSB of system. First, the voltage is increased from 0 V to 60 V in 5 V-steps every 2 s. At 60 V, the voltage remains stable while switching every 2 s in logical order from the 111 to 000 position.

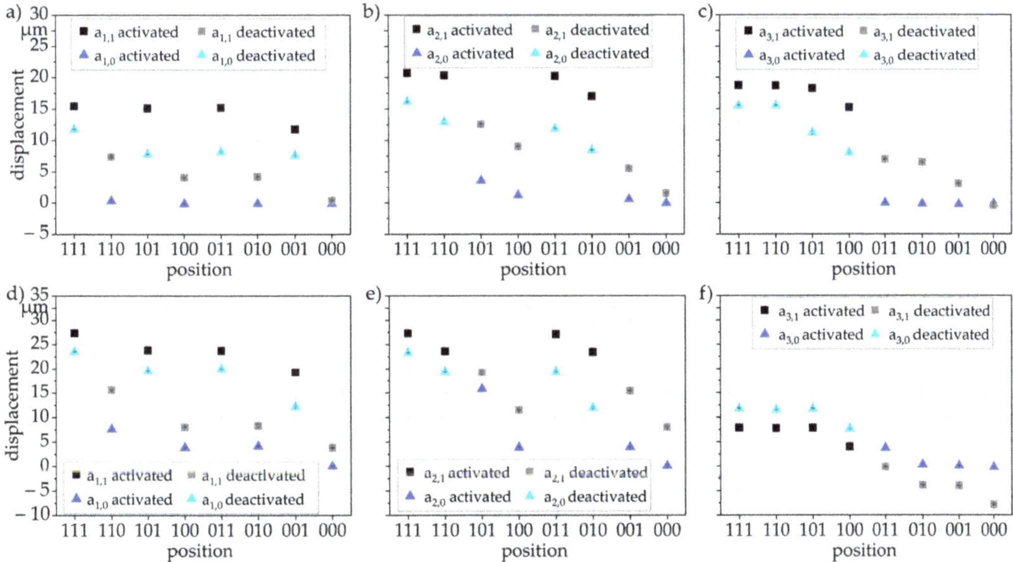

Figure 12. Experimental average actuator displacement of the single bits; each experiment was repeated three times. The experimental conditions correspond to the description in Figure 11. (**a**) Bit 1, (**b**) bit 2 and (**c**) bit 3 of system 1 at 60 V and (**d**) bit 1, (**e**) bit 2 and (**f**) bit 3 of system 2 at 75 V.

In Figure 11, we exemplarily show the time-dependent displacement of the actuators of the LSB. To reach the starting 111 position, the voltage is increased in 5-V-steps every 2 s. At the selected control voltage, the positions are switched in 2-s intervals in logical order from the 111 to 000 position. During the increase of the voltage, clearly visible displacements start at 35 V for the activated actuators $a_{1,1}$ and the deactivated actuators $a_{1,0}$. The displacement of the deactivated actuators is a passive displacement caused by the displacement of the activated actuators. Switching from the 111 to 110 position, the $a_{1,0}$ actuators are activated and the $a_{1,1}$ actuators deactivated.

Based on the experiment presented in Figure 11, the average displacement of each bit is shown in Figure 12. The DAC of system 1 (Figure 12a–c) shows a quite stable behavior. For each bit, the activated actuators shift to approx. the same displacement. The deactivated actuators follow the displacement of all the activated actuators in the microsystem. The shift of the deactivated actuators is less than the shift of the activated actuators.

For system 2, the displacement of the deactivated actuators of bit 3 (Figure 12f) exceeds the displacement of the activated actuators. The very low shift of the MSB is possibly due to the high input stiffness of mechAMP– that bit 3 is connected to, resulting in a less-stable displacement of bit 1 and bit 2, as shown in Figure 12d,e.

4.2.3. Characterization of the Combined DAC and mechAMP

To characterize the system, the single actuators are activated in logical order from the 111 to 000 position, including all the intermediate steps switching after 2 s. The displacements of the DAC and the mechAMP are recorded and analyzed by marking strategic tracking points using video analyses

Compared to the originally simulated systems in Table 2, the displacements differ slightly, which is attributed to two aspects: First, the shift of the fabricated electrostatic actuators differs from the solid-state simulation assumption, and second, due to fabrication tolerances, the input stiffness of the mechAMPs slightly deviates. Therefore, the simulation models for Figure 13 are adapted to the experimental characterization results.

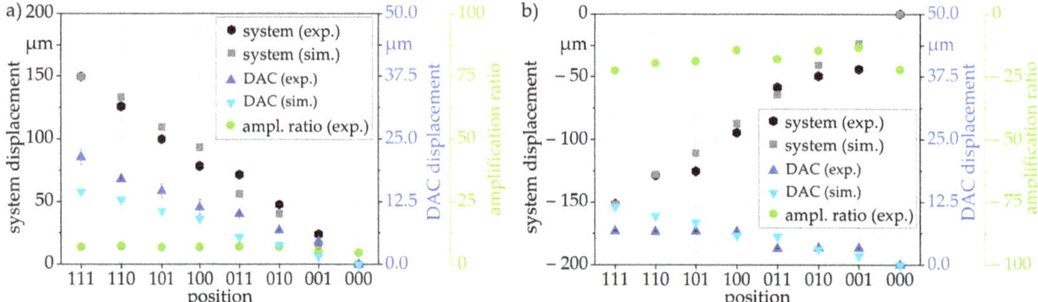

Figure 13. Simulative and experimental displacement and the resulting amplification ratio by displacing the systems in logical order from the 111 to 000 positions, including all the intermediate steps, with a switching time of 2 s for (**a**) system 1 at 60 V and (**b**) system 2 at 75 V (each experiment was repeated three times).

During the experiment, system 1 in Figure 13a shows a total stroke of 149.5 ± 0.3 μm, with an average step size of 21.4 ± 6.1 μm at the control voltage of 60 V. Compared to the simulation, the displacement shows a slight deviation at the 011 and 100 positions. The DAC reaches a total displacement of 21.4 ± 1.5 μm, with an average step size of 3.1 ± 1.1 μm. The amplification ratio calculated by Equation (3) amounts to 6.5 ± 0.9.

The mechAMP− of System 2 (Figure 13b) displaces inversely to the DAC displacement (Figure 6b). Therefore, an upwards displacement of the DAC results in a downwards displacement of mechAMP− and vice versa. Consequently, the total measured stroke at the output of mechAMP− is negative and amounts to -151.3 μm, resulting in an average step size of -21.6 ± 12.2 μm. As the variation in step size is only visible in the experimental results and does not appear during the simulation, the experimental variation is attributed to the fabrication deviations and to the high aspect ratio of the system size to displacement range. The amplification ratio of mechAMP− amounts to -16.9 ± 3.7.

4.2.4. Resonant Behavior, Switching Speed and Reliability of the System

During characterization, a minimum switching time of the bending-plate actuators between two input binary states of 0.1 ms was used successfully. The switching time of the electrostatic actuators is equal to the minimum switching speed of the device. Therefore, the minimum switching time between two input binary states is 0.1 ms, and the system can realize a frequency of up to 10 kHz. Comparing the speed of actuation to other electrostatic-actuated MEMS known from the literature, Grade et al. [17] reported an electrostatic comb-drive actuator system with approx. the same travel range as presented here of 150 μm in less than 1 ms. By characterizing a DAC based on electrostatic parallel-plate actuation, Liao et al. [6] reported a much smaller displacement of 1050 nm in less than 80 ms.

The mechanical system resonance is the limiting factor regarding the system switching frequency due to the oscillating effects. Therefore, the oscillation at the amplifier output and at the DAC was measured. System 1 was switched in logical order with all the intermediate steps from the 111 to 000 position and, finally, from the 000 to 111 position without intermediate steps. The positions were switched in 0.25-s intervals corresponding to a frequency of 4 Hz. The experiment was recorded with 4000 fps.

The oscillation amplitude of the DAC fades faster than the oscillation amplitude at the output of the mechAMP. The maximum oscillation amplitude always exceeds the step size. When switching directly from the 000 to 111 position, the system stroke has a maximum oscillation amplitude of 150.2 μm for the system and 16.6 μm for the DAC. Switching from the 100 to 011 position, the system travels the lowest displacement and exhibits a maximum oscillation amplitude of 9.5 μm for the system and of 0.3 μm for the DAC. Therefore, single-step displacement influences the oscillation amplitude. Additionally, the time until the mechAMP reaches an oscillation amplitude of less than 1 μm depends on the displacement range. The minimum time is 0.08 s for switching from the 100 to 011 position. The maximum time is 0.2 s when switching directly from the 000 to 111 position. Therefore, depending on the required system properties, it can be useful to wait up to 0.2 s until the oscillation of the system output calms down.

The results presented in Sections 4.2.2–4.2.4 are based on experiments that were repeated three times. The deviation of the displacement of system 1 was in the range of 1.5 μm and below for a switching time of 2 s. For examining the switching speed and resonant behavior, the experiment was repeated with a faster switching speed (0.1 ms), resulting in a displacement of 149.7 μm, which is in the range of the results shown in Figure 13a (149.5 ± 0.3 μm). Therefore, a high reliability of the device operation independently from the switching speed is concluded.

4.3. High-Resolution Mode Enlarging the System Positions from $2^3 = 8$ to $3^3 = 27$

In Section 2.1, we described the system function as used in the previous sections. As announced in Section 1, a change in the electrode activation sequence allows to approach twenty-seven discrete positions with the same system. Therefore, a third stable state was added to the single bits. Initially, we only activated either the upwards or the downwards actuators and, thus, reached the known states of "1-direction" or "0-direction", as shown in Figure 14a,b. The intermediate "i-direction" (Figure 14c) is achieved by activating both the upwards and the downwards actuators simultaneously.

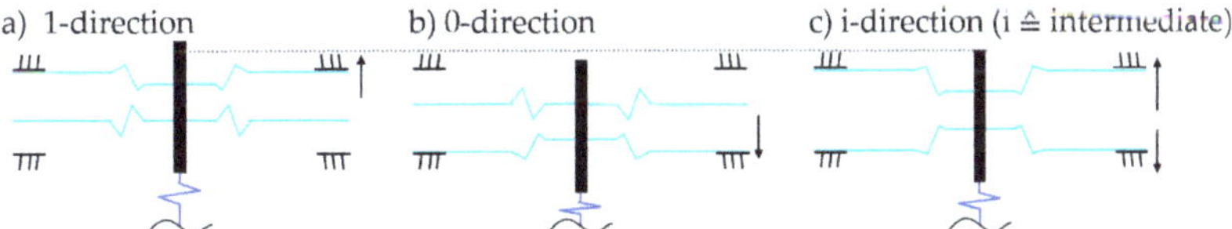

Figure 14. Bit switched to (**a**) 1-direction by activating the $a_{i,1}$ actuator, (**b**) 0-direction by activating the $a_{i,0}$ actuator and (**c**) i-(intermediate) direction by activating both the $a_{i,0}$ and the $a_{i,1}$ actuator.

The i-position does not increase the total system throw but allows to approach a larger number of defined positions and to decrease the step size. Therefore, the number of approachable positions increases from $2^3 = 8$ to $3^3 = 27$ in a 3-tri-state, with each bit having three distinct stable states. When activating the system, including the i-position, Equation (2) and its presumptions are no longer valid. Due to the controlled displacement of the actuators, the i-position is a well-defined stable position. The experimental and simulative approaches show that the actuators have to be activated to reach a well-defined multi-stable intermediate displacement. When neither the 1- nor the 0-direction actuators are activated, an intermediate position is approached by the single bit, but the system output is not defined.

The 3-tri-state activation is performed at 75 V as stiction issues appear at lower voltages due to the increased stiffness of the electrostatic actuators when activating the i-direction. The positions are switched in logical order from the 111 to 000 position, including all the intermediate steps. The experimental and simulative results presented in Figure 15 show that it is possible to realize an approximately linear stepwise displacement with the 3-tri-state system. As the system was never designed for 3-tri-state activation, slight deviations appeared during the simulation and experiment, but nonetheless, the results are promising and show a high potential to reach 27 defined positions with the originally binary eight-positions system.

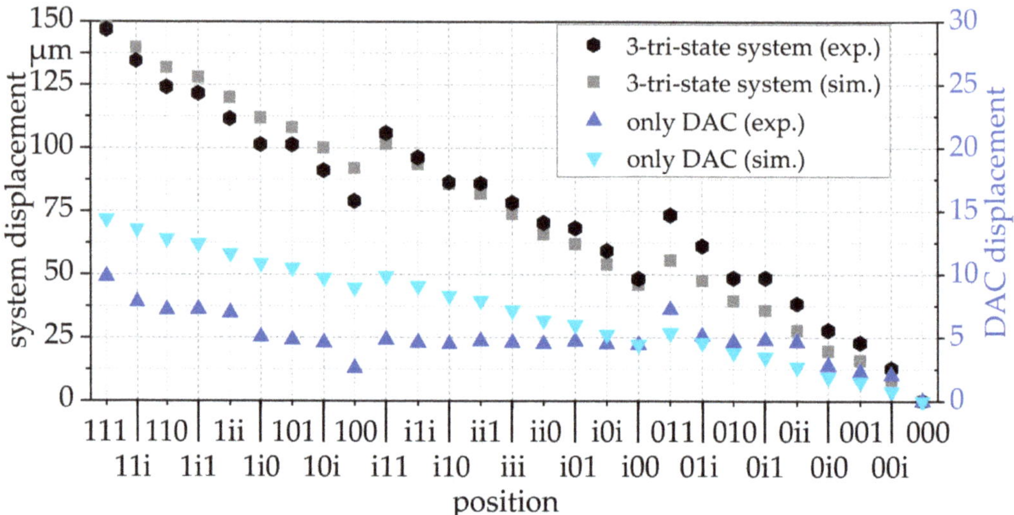

Figure 15. Experimental and simulative displacement of the DAC and the system when activating the microsystem as a 3-tri-state system at the 75-V control voltage. The positions are approached in logical order from the 111 to 000 position including the intermediate i-position (Figure 14c).

5. Conclusions

In this paper, we presented a MEMS-based electromechanical 3-bit DAC system with a maximum stepwise displacement of 149.5 μm at a voltage of 60 V within a minimum switching time of 0.1 ms. The DAC is based on multiple electrostatic bending-plate actuators connected in a parallel configuration enabling a high actuator stiffness and a low pull-in voltage at the same time. The DAC output is directly connected to a mechanical amplifier that enlarges the binary encoded DAC displacement to a large and still binary encoded system displacement.

Based on a derived system transfer function, two different systems were designed. System 1 enlarged the DAC displacement by a factor of 6.5, and System 2 enlarged and inverted the DAC displacement by a factor of -16.9. The experimental results verified the system function, the functionality of the DAC and the combined system, as well as the resonant behavior and the switching speed.

Additionally, an alternative high-resolution mode using three distinct stable states of each bit, resulting in $3^3 = 27$ discrete system positions, was found and experimentally verified. The high-resolution mode does not increase the total stroke, but it allows 27 positions instead of eight binary steps due to a different consecutive sequence of switched electrodes.

Author Contributions: Conceptualization, L.S., P.S. and M.H.; methodology, L.S.; software, L.S. and P.S.; validation, L.S.; formal analysis, L.S.; investigation, L.S.; resources, L.S., P.S. and M.H.; data curation, L.S.; writing—original draft preparation, L.S.; writing—review and editing, P.S. and M.H.; visualization, L.S.; supervision, M.H.; project administration, M.H. and funding acquisition, M.H. All authors have read and agreed to the published version of the manuscript.

Funding: This study was funded by the Deutsche Forschungsgemeinschaft (DFG, German Research Foundation)–Project-ID 287022738–TRR 196, Project C12.

Data Availability Statement: The data can be provided by the author L.S. upon reasonable request.

Acknowledgments: Some segments of the fabrication process were performed at the Center for Interface-Dominated High Performance Materials (ZGH), Ruhr University Bochum.

Conflicts of Interest: The authors declare no conflict of interest. The funders had no role in the design of the study; in the collection, analyses or interpretation of the data; in the writing of the manuscript or in the decision to publish the results.

References

1. Fu, X.; Yang, F.; Liu, C.; Wu, X.; Cui, T.J. Terahertz Beam Steering Technologies: From Phased Arrays to Field-Programmable Metasurfaces. *Adv. Opt. Mater.* **2020**, *8*, 1900628. [CrossRef]
2. Schmitt, L.; Schmitt, P.; Liu, X.; Czylwik, A.; Hoffmann, M. Micromechanical Reflect-Array for THz Radar Beam Steering based on a Mechanical D/A Converter and a Mechanical Amplifier. In Proceedings of the 2020 Third International Workshop on Mobile Terahertz Systems (IWMTS), Essen, Germany, 1–2 July 2020.
3. Liu, X.; Samfaß, L.; Kolpatzeck, K.; Häring, L.; Balzer, J.C.; Hoffmann, M.; Czylwik, A. Terahertz Beam Steering Concept Based on a MEMS-Reconfigurable Reflection Grating. *Sensors* **2020**, *20*, 2874. [CrossRef] [PubMed]
4. Sarajlic, E.; Collard, D.; Toshiyoshi, H.; Fujita, H. 12-bit microelectromechanical digital-to-analog converter of displacement: Design, fabrication and characterization. In Proceedings of the 2007 IEEE 20th International Conference on Micro Electro Mechanical Systems (MEMS), Hyogo, Japan, 21–25 January 2007.
5. Zhou, G.; Logeeswaran, V.J.; Fook, S.C. An open-loop nano-positioning micromechanical digital-to-analog converter for grating light modulation. *IEEE Photonics Technol. Lett.* **2005**, *17*, 1010–1012. [CrossRef]
6. Liao, H.-H.; Yang, Y.-J. A micromirror module using a MEMS digital-to-analog converter and its application for optical surface profiling. *J. Micromech. Microeng.* **2010**, *20*, 105009. [CrossRef]
7. Richard, Y.; Conant, R.A.; Pister, K.S.J. Mechanical digital-to-analog converters. In Proceedings of the Tenth International Solid-State Sensors and Actuators Conference, Sendai, Japan, 7–10 June 1999.
8. Liu, Q.; Huang, Q.-A. Design and finite element analysis of weighted-stiffness microelectromechanical digital-to-analogue converters. *Electron. Lett.* **2001**, *37*, 755–756. [CrossRef]
9. Pandiyan, P.; Uma, G.; Umapathy, M. Design and simulation of MEMS-based digital-to-analog converters for in-plane actuation. *Arab. J. Sci. Eng.* **2017**, *42*, 4991–5001. [CrossRef]
10. Toshiyoshi, H.; Kobayashi, D.; Mita, M.; Hashiguchi, G.; Fujita, H.; Endo, J.; Wada, Y. Microelectromechanical digital-to-analog converters of displacement for step motion actuators. *J. Microelectromech. Syst.* **2000**, *9*, 218–225. [CrossRef]
11. Non-Member, E.S.; Collard, D.; Toshiyoshi, H.; Fujita, H. Design and modeling of compliant micromechanism for mechanical digital-to-analog conversion of displacement. *IEEJ Trans. Electr. Electron. Eng.* **2007**, *2*, 357–364.
12. Schmitt, P.; Hoffmann, M. Engineering a compliant mechanical amplifier for MEMS sensor applications. *J. Microelectromech. Syst.* **2020**, *29*, 214–227. [CrossRef]
13. Martin, H.; Nüsse, D.; Voges, E. Electrostatic parallel-plate actuators with large deflections for use in optical moving-fibre switches. *J. Micromech. Microeng.* **2001**, *11*, 323.
14. Schmitt, P.; Schmitt, L.; Tsivin, N.; Hoffmann, M. Highly Selective Guiding Springs for Large Displacements in Surface MEMS. *J. Microelectromech. Syst.* **2021**, *30*, 597–611.
15. Schmitt, L.; Liu, X.; Czylwik, A.; Hoffmann, M. Design and fabrication of MEMS reflectors for THz reflect-arrays. In Proceedings of the 2021 Fourth International Workshop on Mobile Terahertz Systems (IWMTS), Duisburg, Germany, 5–7 July 2021.
16. Ibrahim, S.; Zeimpekis, I.; Kraft, M. A dicing free SOI process for MEMS devices. *Microelectron. Eng.* **2012**, *95*, 121–129.
17. Grade, J.D.; Jerman, H.; Kenny, T.W. Design of large deflection electrostatic actuators. *J. Microelectromech. Syst.* **2003**, *12*, 335–343. [CrossRef]

Article

Investigation of the Dynamics of a 2-DoF Actuation Unit Cell for a Cooperative Electrostatic Actuation System

Almothana Albukhari [1,2,*] **and Ulrich Mescheder** [1,3]

1. Mechanical and Medical Engineering Faculty, Institute for Microsystems Technology (iMST), Furtwangen University, 78120 Furtwangen, Germany; mes@hs-furtwangen.de
2. Department of Microsystems Engineering (IMTEK), University of Freiburg, 79110 Freiburg, Germany
3. Associated to the Faculty of Engineering, University of Freiburg, 79110 Freiburg, Germany
* Correspondence: alb@hs-furtwangen.de

Abstract: The mechanism of the inchworm motor, which overcomes the intrinsic displacement and force limitations of MEMS electrostatic actuators, has undergone constant development in the past few decades. In this work, the electrostatic actuation unit cell (AUC) that is designed to cooperate with many other counterparts in a novel concept of a modular-like cooperative actuator system is examined. First, the cooperative system is briefly discussed. A simplified analytical model of the AUC, which is a 2-Degree-of-Freedom (2-DoF) gap-closing actuator (GCA), is presented, taking into account the major source of dissipation in the system, the squeeze-film damping (SQFD). Then, the results of a series of coupled-field numerical simulation studies by the Finite Element Method (FEM) on parameterized models of the AUC are shown, whereby sensible comparisons with available analytical models from the literature are made. The numerical simulations that focused on the dynamic behavior of the AUC highlighted the substantial influence of the SQFD on the pull-in and pull-out times, and revealed how these performance characteristics are considerably determined by the structure's height. It was found that the pull-out time is the critical parameter for the dynamic behavior of the AUC, and that a larger damping profile significantly shortens the actuator cycle time as a consequence.

Keywords: cooperative actuators; inchworm motor; electrostatic actuator; MEMS; FEM; coupled-field modeling; gap-closing actuator; pull-in time; pull-out time; squeeze-film damping

Citation: Albukhari, A.; Mescheder, U. Investigation of the Dynamics of a 2-DoF Actuation Unit Cell for a Cooperative Electrostatic Actuation System. *Actuators* **2021**, *10*, 276. https://doi.org/10.3390/act10100276

Academic Editors: Manfred Kohl, Stefan Seelecke and Ulrike Wallrabe

Received: 1 August 2021
Accepted: 5 October 2021
Published: 18 October 2021

1. Introduction

Electrostatic actuation is widely commercially used in applications where only low forces and displacements are needed, such as in MEMS-based gyros using the Coriolis effect [1] and in the Digital Micro Mirror (DMD) device [2]. The growing interest in electrostatic actuators observed in recent decades in commercial and research circles is owed primarily to their ease of fabrication and integration due to the compatibility with CMOS fabrication processes, as well as their outstanding force-to-volume ratio, which favors them among other actuation means in the micrometer range. Additional advantages of electrostatic actuators include their low energy consumption, relatively fast response and high resonance frequencies, and low temperature dependence [3,4]. However, electrostatic actuators in MEMS have intrinsic limitations in terms of force and linear displacement capacities. To overcome these limitations, electrostatically driven actuation systems based on the inchworm principle, which typically achieves extended linear displacements through a scheme of repeated latch-drive operations on a common shuttle, have been proposed [5–8]. However, even when using the so-called gap-closing mechanism and cooperative action of different actuators, the reported forces are rather small and typically measure less than 1 mN. Recently, inchworm motors have been integrated in more complex schemes, such as a six-legged terrestrial robot [9], and this year, a robust electrostatic inchworm motor has been presented, allowing 80 mm stroke and a force of 15 mN at 100 V driving voltage.

Thus, the authors claimed a performance of their electrostatic inchworm motor similar to piezoelectric-driven inchworm motors [10].

In principle, an electrostatic gap-closing actuator (GCA) can provide large forces. However, a great influence on the dynamic behavior of this kind of actuator, which requires a thorough consideration, arises from the pull-in phenomenon, also known as the pull-in instability. The distance at which the pull-in takes place in a GCA, known as the pull-in distance, is one-third of the nominal gap when a linear force flexure is used, but it can be extended by using higher-order force flexures [11,12]. In 2007, Zhang et al. [3] surveyed the pull-in phenomenon in addition to key issues concerning the stability, nonlinearity, and reliability of electrostatic MEMS actuators. Then in 2014, Zhang et al. [13] contributed a more extensive review on pull-in instability, in which they covered the theoretical background of the phenomenon, many of the aspects it influences and is influenced by, and its various applications.

Moreover, depending on the application, achieving a decent grasp of the mechanical dynamic behavior of a MEMS actuator, such that its design and operation goals are secured, requires taking into account not only the relevant stiffness and inertia elements but also the damping mechanisms of energy dissipation affecting the structure. These damping mechanisms are usually classified into extrinsic and intrinsic mechanisms. In MEMS applications where special measures are not taken to eliminate extrinsic losses, e.g., to achieve very high quality factors in resonators, the intrinsic damping sources such as thermoelastic damping, for instance, are almost negligible when compared to extrinsic types of dissipation, especially in single-crystalline materials. Moreover, extrinsic losses, which arise from the structure's interaction with its environment, can be dominating and greatly influence its dynamic behavior. Consequently, it has been commonly reported that the most significant damping source in MEMS is viscous or air damping [14] (pp. 97–114). Viscous damping can take up to three forms in MEMS: drag, shear-film, and squeeze-film damping (SQFD). The first one relates to the friction experienced by a structure moving in free space in air, whereas the other two relate to structures moving in close proximity to each other. Shear-film damping happens when structures move parallel to one another and their relative distance remains constant, whereas SQFD happens when they come closer to one another during movement, i.e., when the gap between them diminishes and the fluid therein is squeezed out. In MEMS structures featuring large flat surfaces that move across a gap between them, which is much smaller than their lateral dimensions, and where a thin film of fluid resides, the SQFD dominates, and in many cases it is the only source of damping granted attention [14] (p. 118).

Taking into account the aspects discussed so far, it can be easily understood why the modeling and simulation of MEMS devices is one of the most demanding tasks, especially if the device is intended to have a certain dynamic performance. Although, from a structural point of view, a MEMS component can sometimes be relatively simple; however, the complexity of its modeling and the difficulty of gaining a thorough understanding and accurate prediction of its dynamic behavior stems from the fact that it usually entails the inherent coupling of various energy domains and physical forces (mechanical, electrical, fluidic, etc.). The interactions among these domains are in many cases intrinsically nonlinear, especially for large-amplitude motions of microstructures that amplify their geometric nonlinearities as well [15] (pp. 7–8). This reality makes FEM tools essential for the design and development of novel MEMS applications [14] (pp. 4–11), such as the one at hand. With FEM, small deviations from the ideal structures can also be considered, such as not perfectly vertical sidewalls or small geometrical variations, which occur due to the limitations of the microfabrication processes and can have a large impact on the system behavior of such a complex cooperating system.

This paper presents the results of modeling an actuation unit cell (AUC), both analytically by a lumped-parameter model and numerically with FEM, and compares between the two approaches where possible. The AUC constitutes the basic building block in a novel concept of a cooperative electrostatic microactuator system that is intended for realizing large actuation displacements (tens of mm) and forces (tens of N) [16], which can meet the requirements of implantable actuators for bone distraction in osteogenesis applications, for example [17]. Therefore, the concept of the actuation system as a whole and the main cooperative operations and mechanisms therein will be briefly discussed. Furthermore, a simplified analytical model of the AUC is presented, followed by FEM simulations with several types of studies carried out on a parameterized model to reveal different aspects of the dynamic behavior of the actuator and its operational parameters of interest, e.g., the static study of the deflection-voltage curve that identifies the pull-in and pull-out transitional points and the transient studies of the actuation and settling times, which make up the major parts of the actuation cycle. Moreover, the parameters investigated by FEM are compared with available analytical models found in the literature. The modeling of viscous damping, namely the squeeze-film damping, and its influences on the dynamic behavior are discussed as well.

2. System Concept, Materials, and Methods

2.1. The Cooperative Concept and Actuation Unit Cell

The works presented here are a stepping stone to implement a modular-like concept of a cooperative electrostatic microactuator system. The cooperative system as a whole is meant to provide scalable displacement in the millimeter range and force capacities of several newtons. A sketch of the system concept is shown in Figure 1. As shown in the figure, pairs of AUCs are integrated with another multistable actuator subsystem, which consists of two sliding shafts that are mounted on central mechanical tracks. The two shafts are initially interdigitated with a stationary holder via mechanical suspension springs that pull them together. Mobility transmission electrodes placed parallel to the sliding shafts provide electrostatic forces to pull out the shafts and release them from the stationary holder such that the outer shafts' teeth become clutched by those of the AUCs. When the shaft is mobile, linear movement is realized by actuating the AUCs in a scheme that is detailed below. However, it is worth noting that the system model is based upon the cooperation of an AUC with many other counterparts to actively actuate the shaft in a bidirectional manner. Therefore, the AUC itself is a two-degree-of-freedom (2-DoF) actuator that is able to both clutch with the main shaft as well as actuate it in either one of the two directions along the shaft.

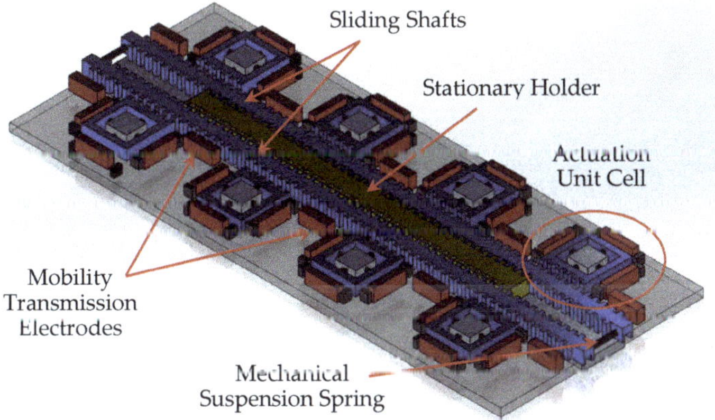

Figure 1. System concept of a multistage, multistable inchworm motor, illustrated by four pairs of actuation unit cells (AUC) distributed at both sides of two connected sliding shafts (long blue structures), which are pulled together by mechanical suspension springs and held at idle states by a central stationary holder (yellow). The mobility transmission electrodes determine whether the shafts are interlocked with the stationary holder or released from it.

The proposed cooperative approach aims at combining the efforts of multiple AUCs to overcome the intrinsic limitations of each one, namely, in terms of the delivered force and displacement. Pushing the limit of the actuation force is achieved by actuating multiple AUCs simultaneously (parallel operation; timewise), whereas exceeding the displacement capability is realized by the inchworm mechanism, i.e., by activating multiple AUCs sequentially and repeatedly to displace the shaft (serial operation; timewise). Moreover, on the one hand, as it can be inferred from Figure 1, the shaft has multiple stable positions, the distance between which will be determined by the technological manufacturing limitations and design specifications of the interlocking teeth of the stationary holder and the sliding shafts; on the other hand, the displacement of the AUC, i.e., the stroke it is capable of, must be kept small to generate sufficiently large forces, as discussed later. As a result, it must be taken into account that a step displacement of the shaft between two neighboring stable positions (pitch of the interlocking teeth) might require multiple AUC strokes provided by multiple groups of AUCs. Figure 2 illustrates an example of the proposed cooperative scheme with a simple diagram in which each of the units, labelled A, B, and C, represents a group of AUCs that are driven in parallel. Therefore, in this illustration, the AUCs are driven in three groups to satisfy the presumed requirement that a shaft step requires three AUC strokes.

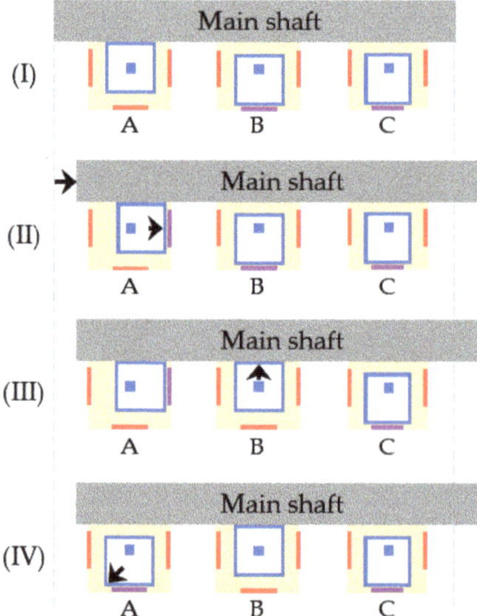

Figure 2. A diagram demonstrating the scheme of serial operations of the unit cells to actuate the main shaft. Each of the units A, B, and C represents a parallel-driven group of AUCs. In this illustration, a step displacement of the shaft (a full actuation cycle) requires three successive strokes by three AUC groups. The diagram reflects the operation of a single group, A, and how it overlaps with the other groups in a cycle. The small and large blue squares represent the anchor and moveable electrode in an AUC, respectively. The purple or red strips represent the electrically charged or grounded electrodes, respectively. It starts after the main shaft has been made mobile by disengaging it from the stationary holder (yellow structure in Figure 1). Simultaneously, the shaft will be clutched by group A, whose teeth are aligned to those of the shaft, whereas the other groups are already pulled away (**I**). Group A then actuates the shaft (side actuation; rightward arrows) until reaching the stroke limit (**II**). Afterwards, group B, which is now aligned to the shaft, returns to home position (upward arrow) and clutches the shaft (**III**). Lastly, group A disengages from the shaft (diagonal arrow) to allow group B to handle the shaft from this point onward (**IV**), then similarly hand it over to group C (not shown), and the cycle repeats.

The mechanical structure proposed for the AUC has a rigid, rectangular-shaped, centrally anchored, and electrically grounded movable electrode (see Figure 3). The elastic restoration force is realized by four serpentine flexures that are arranged symmetrically about the centered anchor. The frame is actuated via electrostatic attraction towards either one of three surrounding stationary electrodes. Electrode 3 determines whether the frame is engaged with the shaft or not, whereas the two Electrodes 1 and 2 will move it in either of the two opposite directions. The mechanical stoppers prevent the electrodes from coming into contact. Table 1 lists the values for the dimensions corresponding to Figure 3.

2.2. The Analytical Model of the AUC

The AUC is basically a spring–mass–damper system that is driven by external electrostatic force. As it can be inferred from Figure 3, the moveable electrode has a planar movement in the xy-plane; however, if one considers its motion along the x-axis, an approximation can be made by viewing it as a single-degree-of-freedom device that is confined to the movement along that axis. Additionally, due to the geometrical and operational similarity between the two primary axes of motion, investigating the behavior of the system along one of them essentially covers both. Consequently, as shown in Figure 4, the parallel-plate electrostatic actuator is adopted as an approximate lumped-parameter model, which assumes the infinite rigidity of the moveable electrode wall and that its outer vertical surfaces remain parallel to the opposite surfaces of the stationary electrodes during motion.

Figure 4 shows the four forcing elements that primarily govern the dynamic behavior of the system: The first is the electrostatic force exerted on the grounded moveable electrode (Moveable Plate) when a stationary electrode (Fixed Plate 1) is charged with an electric potential V_0. In line with the concept of operation discussed above, only one of the two electrodes actuating the structure in the x-axis may be charged at any given time. The second is the inertial element that consists of the mass of the moveable electrode wall and a portion of the serpentine springs' masses. It is represented by an equivalent mass for the moveable plate, $m_{eq} = m_{wall} + r\, m_{springs}$, where r is the springs' effective mass ratio. The third element is the elastic force resulting from an equivalent spring constant k_{eq} that reflects the overall stiffness of the serpentine flexures along the x-axis. The equivalent spring constant was derived based on energy methods, as shown in Appendix A. The last element is the viscous damping force represented solely by the dominant SQFD that affects the moveable electrode on both sides as it moves along the x-axis. Modeling the latter is explained in more detail below.

Accounting for the viscous damping of fluids in MEMS is fundamentally based on the Navier–Stokes equation, which is a nonlinear three-dimensional partial differential equation that is very challenging to solve numerically, let alone analytically [14] (p. 114). Consequently, certain simplifications and assumptions were customarily made by researchers, depending on specific aspects of the application under investigation, to allow this problem to be tackled by a simpler nonlinear or even a linearized Reynolds equation. Good reviews of some of these efforts in the MEMS field can be found in [14,18].

The fluidic thin films occupying the spaces between the vertical, parallel, and relatively large side walls of the electrodes (the moveable electrode with respect to each of the stationary electrodes) contribute to most of the damping in this application. Moreover, the squeeze-film effect, which results from the moveable electrode's gap-closing motion, may exert a damping force or a spring force or both. The proportionality and extent of these forces depends on several factors that are summarized in the nondimensional so-called squeeze number, which serves as a measure of the compressibility of the thin film. For two parallel plates with overlapping area A, separated by distance d, and with a harmonic excitation of one plate at frequency ω, the squeeze number σ is obtained as follows:

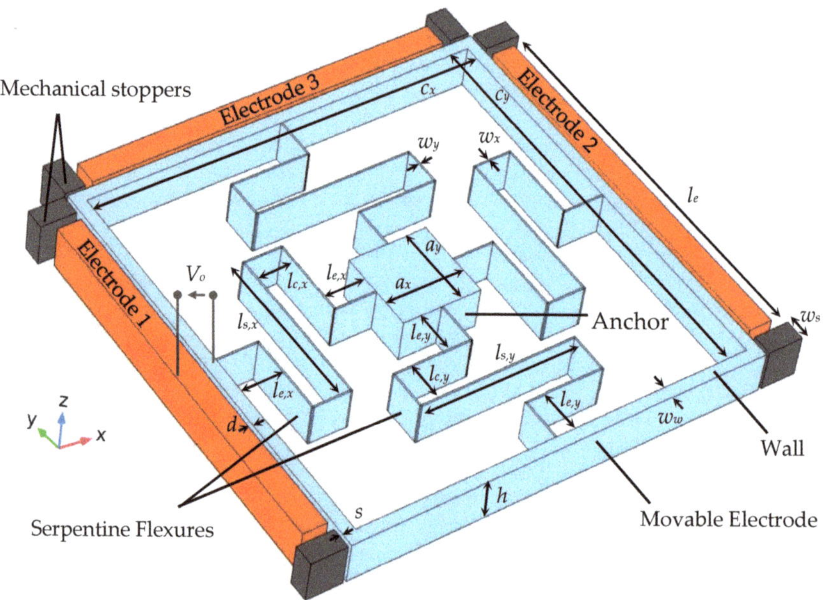

Figure 3. Three-dimensional schematic of the proposed AUC for the cooperative microactuator system, taken from the FEM model in COMSOL.

Table 1. Typical parameter values for the AUC in the FEM model.

Description	Symbol	Value	Unit
Height of unit cell	h	50	μm
Width of cell wall	w_w	15	μm
Size of cell wall along x-axis (inner dimension)	c_x	500	μm
Size of cell wall along y-axis (inner dimension)	c_y	500	μm
Size of anchor (x-axis)	a_x	100	μm
Size of anchor (y-axis)	a_y	100	μm
Width of x-axis spring (flexure width)	w_x	2	μm
Width of y-axis spring (flexure width)	w_y	2	μm
Length of connector beams of x-axis spring	$l_{c,x}$	50	μm
Length of span beam of x-axis spring	$l_{s,x}$	200	μm
Length of extension beams of x-axis spring	$l_{e,x}$	53	μm
Length of connector beams of y-axis spring	$l_{c,y}$	50	μm
Length of span beam of y-axis spring	$l_{s,y}$	200	μm
Length of extension beams of y-axis spring	$l_{e,y}$	53	μm
Length of stationary electrodes	l_e	460	μm
Width of stopper (length of contact area)	w_s	25	μm
Nominal air gap between electrodes	d	5	μm
Stroke (distance between mov. electrode at rest and stopper)	s	4	μm
Young's modulus	E	170	GPa
Poisson's ratio	ν	0.28	-
Density (c-Si)	ρ	2329	kg/m^3
Air viscosity	μ	1.845×10^{-5}	Pa.s
Actuation Voltage	V_o	variable	V

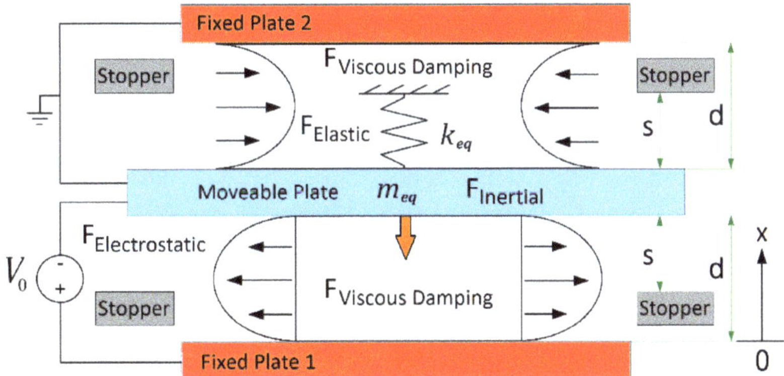

Figure 4. Lumped-parameter modeling of the AUC as a parallel-plate capacitor with squeeze-film damping. m_{eq} is the moveable plate's equivalent mass, k_{eq} is the serpentine flexures' equivalent spring constant, x is the position of the moveable plate, d is the initial (nominal) gap, s is the stroke extending to the stoppers, and V_0 is the actuation voltage.

$$\sigma = \frac{12\, A\omega\mu}{P_a d^2},\tag{1}$$

where μ and P_a are the fluid viscosity and ambient pressure of the thin film, respectively [14] (p. 128). A low squeeze number means that the gas does not undergo compression; hence, the assumption of incompressible gas is more acceptable, and its damping force dominates with a negligible spring effect. On the other hand, a high squeeze number means that the fluid is trapped in the gap by its own viscosity and cannot escape easily; thus, it undergoes compression and its spring effect becomes more significant to the extent that it acts as a spring rather than a damper with much larger squeeze numbers [19]. The transition point at which the damping and stiffness forces of the film become equal has been worked out through the so-called cut-off squeeze number σ_c. Blech [20] derived an analytical expression for it as follows:

$$\sigma_c = \pi^2 \left(1 + (w/l)^2\right),\tag{2}$$

where w and l are the width and length of the thin film that correspond to h and l_e of the AUC, respectively. The geometrical dimensions defined in Table 1 result in $\sigma_c \approx 9.97$. Under the assumptions of operation at atmospheric pressure, almost free air access to the gap from its surroundings, and moderate driving frequencies, squeeze numbers much less than unity are expected, i.e., $\sigma \ll \sigma_c$. Hence, the incompressible fluid assumption is made in the analysis presented here, and the spring effect of the SQFD is neglected.

Furthermore, under assumptions satisfying the Reynolds equation [14] (p. 122) and the incompressible-gas approximation, Starr [21], based on the assumptions of small deflection and small pressure variation, used a linearized form of the Reynolds equation to derive the damping factor for a rectangular plate, which can be expressed as follows:

$$c = \frac{w^3 l\, \mu}{d^3} \cdot f(w/l),\tag{3}$$

where $f(w/l)$ is a factor that depends on the aspect ratio of the thin film, and it is obtained, as reported in [22], by the following:

$$f(w/l) = 1 - 0.6\,(w/l), \quad for \ \ 0 < w/l < 1.\tag{4}$$

Contrary to Starr's assumptions, the proposed AUC's moveable electrode experiences large deflections that give rise to large pressure variations. To this end, the approaches adopted in [5,22] showed that the nonlinear behavior of the SQFD can still be approximated using Starr's expression by replacing the nominal film thickness d with the actual, variable thickness of the film instead. As a result, the viscous damping force exerted by the two thin films in our model can be expressed using the parameter notation of Table 1 and Figure 4 as follows:

$$F_d = c_{SQFD}\dot{x} = \left(1 - 0.6\frac{h}{l_e}\right)h^3 l_e \mu \left(\frac{1}{x^3} + \frac{1}{(2d-x)^3}\right)\dot{x}. \tag{5}$$

In summary, considering the four primary forces in the system discussed here, the equation of motion of the system, with reference to Figure 4, is written as follows:

$$m_{eq}\ddot{x} + c_{SQFD}\dot{x} + k_{eq}(x-d) + \frac{\epsilon_0 A}{2x^2}V_0^2 = 0, \quad for \quad d - s < x < d + s, \tag{6}$$

where ϵ_0 is the vacuum permittivity and $A = hl_e$ is the overlapping electrode surface area.

2.3. FE Modeling and Simulation

The device investigated here is a pull-in type device, similar to a microswitch. Pull-in devices undergo a sharp transition under deflection through a so-called pull-in point until they land on the opposite electrode unless a mechanical stopper prevents this. As such, the numerical modeling presented here aimed at establishing the device's operation around this transitional point, through the static deflection-voltage curve, to determine the primary pull-in characteristics. Additionally, by performing transient analyses, the dynamic behavior was examined in order to determine other important performance characteristics, such as the pull-in and pull-out times, which help to define a stable operation frequency range for the device.

To achieve these objectives, the AUC was modeled in the commercial FE modeling software COMSOL Multiphysics. A 3D model with typical parameter values was used as shown in Figure 3 and Table 1. The model was simulated through a predefined multi-physics interface, the electromechanics, which is part of the structural mechanics module in COMSOL, and combines the solid mechanics and electrostatics interfaces with a moving mesh functionality. The latter was used to mesh the highly deformable air gap, which only occupied the overlapping area between the surfaces of the moveable electrode and Electrode 1. Thus, the fringing fields were neglected, similarly to the analytical model. The actuating electric potential was applied directly across the gap by defining electric potentials V_0 and 0 V to the gap-bordering surface of Electrode 1 and to the bulk of the moveable electrode, respectively. The electrostatics interface computed the electric field in the air gap based on Gauss's law and updated it continuously as the gap geometry changed. The electrostatic force that acts on the moveable electrode was calculated in the multiphysics interface by Maxwell's stress tensor. The displacements, stresses, and strains were solved in the solid mechanics interface based on Navier's equations. Additionally, the fluid thin-film damping could be calculated via an ad hoc boundary condition (BC) predefined in the solid mechanics interface. A 2D model that is obtained by a cross-sectional cut along the z-axis of the 3D model was primarily utilized, which lessened the computational load significantly and made it easier to obtain converging solutions. For the different COMSOL simulations, fully coupled solvers were utilized with relative tolerance values of no more than 0.001.

2.3.1. Static Analysis

The static deflection-voltage curve of the AUC is defined by the equilibrium points at which the stiffness and electrostatic forces are equal. Figure 5 depicts this curve for the device with the nominal dimensions listed in Table 1. The plot was generated by sweeping the deflection of the center point at the movable electrode's surface, which is opposite to the actuating stationary electrode. The deflection sweep was made over the complete possible range of motion of the center point from the undeflected position to the stationary electrode, while the actuating voltage level required for any deflection point was calculated inversely. From the plot, it can be inferred that there are stable and unstable regions depending on whether the device can in reality be driven at those points in a controllable manner or not. Starting from the undeflected position ($V_0 = 0$ V), as the actuation voltage increases, the AUC may be driven in a stable manner up to the point where pull-in takes place, and then it is pulled through the gap and lands on the stoppers (the second stable region, above the stoppers' mark). If the actuation voltage is increased beyond a certain level, the movable electrode wall will collapse under the increasing electrostatic force and come into contact with the stationary electrode. This is essentially a second pull-in point, where the electrostatic force is acting additionally against the wall stiffness; however, considering the operational point of view of the device as a whole, it is referred to as the collapsing point. The amount of deflection above the stoppers line with respect to it represents the static internal deflection, which the wall undergoes at pull-in. Contrariwise, when at the pulled-in position, as the actuation voltage drops below a certain value, the recovering elastic force of the flexures overcomes the electrostatic force and pulls the frame out (pull-out) In this analysis, the 2D model was used.

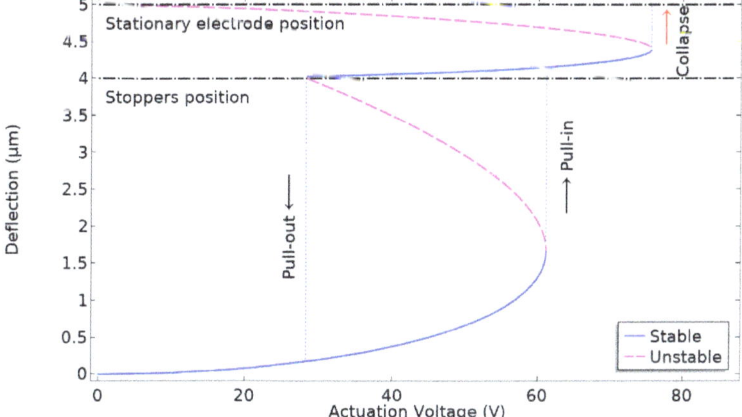

Figure 5. The static deflection-voltage curve for a 2D model of the AUC corresponding to Table 1 generated by simulation with COMSOL. It shows the stable and unstable regions of the curve as well as the major transitional points, namely, the pull-in, pull-out, and collapse transitions. The dot-dashed lines represent the positions of the stoppers and the stationary electrode.

2.3.2. Transient Analysis

The transient analysis is a time-dependent study that was used to investigate the dynamic behavior of the movable electrode under the influence of an actuating electrostatic force of a certain temporal shape or after releasing the structure from the influence of such force. Therefore, it was primarily used to find out the pull-in and pull-out times of the AUC and how these varied with the structures' dimensions and the adopted viscous damping approximations. Moreover, modeling the SQFD effect in the COMSOL environment through the built-in BC available in the solid mechanics interface was only feasible in a 3D model because the borders of the film in the out-of-plane direction could not be properly defined in the 2D model through the available interface settings. As a result, the derived analytical approximation for the SQFD (discussed in Section 2.2) was applied in the 2D model, which is the main model used in transient analyses, as a boundary load acting on the AUC's surfaces of interest. A comparative analysis between the built-in BC and the embedded analytical model in a 3D environment is presented at the end of the discussion chapter.

2.4. Solving the Analytical (Lumped-Parameter) Model by MATLAB/Simulink

To corroborate the findings of the transient analyses obtained by the FEM simulation, the system equation of motion, Equation (6), was implemented in a MATLAB/Simulink model (see Figure 6), which was set up with an ode8 Dormand-Prince solver and a fixed-step of 1 ns. The model was solved with two different sets of initial conditions to demonstrate the pull-in and pull-out behaviors. For the pull-in response, the initial position x_0 was set to d, and V_0 was set to the amplitude of the applied voltage of the corresponding COMSOL simulation to be compared with. The pull-in time was noted when the position x reached the stoppers ($x = d - s$). For the pull-out response, the initial conditions were $x_0 = d - s$ and $V_0 = 0$ V. The parameters of the model, e.g., the electrode surface area and equivalent mass, were calculated based on the geometry through a script; however, to obtain results that are more comparable to the numerical model, the spring constants were extracted from a 3D COMSOL model rather than the analytical formulations. Additionally, it was assumed that the springs' effective mass ratio $r = 0.5$.

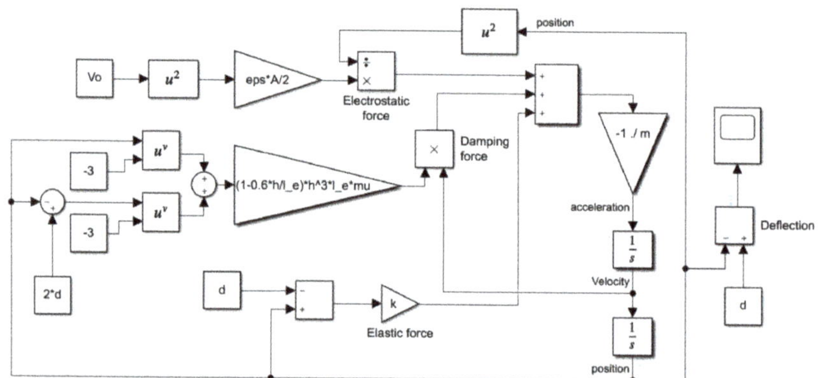

Figure 6. The MATLAB/Simulink model corresponding to the system equation of motion, Equation (6).

3. Results

In the following presentation of FEM simulation results and the parametric sweeps therein, unless otherwise specified, the dimensions listed in Table 1 apply.

3.1. Pull-In Voltage

The pull-in voltage for parallel-plate electrostatic actuators with linear stiffness elements, as shown in Figure 4, is given in [14] (p. 78) as follows:

$$V_{Pull\text{-}in} = \sqrt{\frac{8kd^3}{27\epsilon A}} .$$

(7)

With FEM, different static parametric studies for the AUC's dimensions were simulated. A primary example is the influence of the spring constant k, which was examined through a sweep of the flexure width. Accordingly, Figure 7a plots the numerically and analytically calculated pull-in voltage as a function of the flexure width. The analytical solution determined the spring constant through the formulations given in Appendix A. The difference between the two solutions ranges between 7% and 7.6%. Similarly, a parametric study sweeping the gap width d is presented in Figure 7b. The relative difference between the numerical and analytical solutions here is between 7.7% and 9.3%. The curves in Figure 7 represent a dependance of the pull-in voltage on the square root of the cube of the respective parameter. For ideal structures considered in the scope of these analyses, the reported differences between the FEM and analytical models in the investigated ranges stem mainly from the underestimation of the spring constant by the approximate analytical formulations (see Appendix A). The FEM simulations showed that the pull-in distances, as expected from linear stiffness elements, remained equal to one-third of the nominal gap widths.

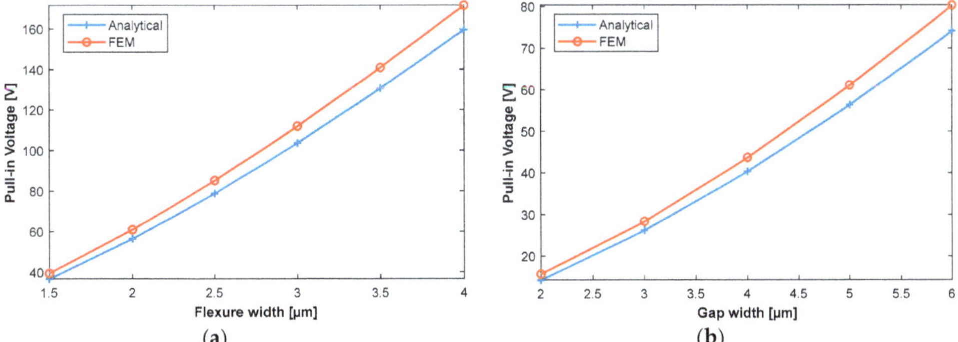

Figure 7. The results of static parametric analyses of the pull-in voltage, where the latter is plotted as a function of the following: (**a**) the flexure width (by sweeping both w_x and w_y); and (**b**) the nominal gap width d. In the plots, the FEM results are compared with the analytical solutions that are based on Equation (7) and the stiffness formulations of the serpentine flexures presented in Appendix A.

3.2. Pull-In Time

The pull-in time is a fundamental parameter for the dynamic behavior of a GCA. Therefore, attempts to derive an analytical equation for it are found in the literature. Naturally, the derivation depends on the proper assumptions and approximations that can be made in the solution of the corresponding system equation of motion. To this end, for a beam MEMS switch, with the assumption of an inertia-limited system and the approximations of a small damping coefficient ($c \cong 0$) and a constant electrostatic force, a closed-form solution of pull-in time was derived in [23] (p. 68) as follows:

$$t_{Pull-in} \cong 3.67 \frac{V_{Pull-in}}{V_o \omega_o} ,\qquad(8)$$

which shows the pull-in time dependence on the ratio of the pull-in voltage $V_{Pull-in}$ relative to the actuation voltage V_o as well as the stiffness and inertia of the structure (lumped in the natural frequency, $\omega_o = \sqrt{k_{eq}/m_{eq}}$). From the surveyed literature, this formulation was deemed the most suitable to the system at hand in terms of the assumptions made to derive it, despite the fact that it lacks the consideration of damping, the influence of which is examined below.

Accordingly, several parametric transient studies were carried out in a 2D model in COMSOL in order to establish how relatable the behaviour of the system at hand is to Equation (8) and how well the assumption of an inertia-limited system is fulfilled. Each parametric study was carried out twice under two different boundary conditions: (a) neglecting the damping of the thin film (undamped) and (b) with the SQFD force exerted as defined in Equation (5) (damped). Moreover, unless otherwise specified, the actuation voltage was applied in the form of a step function with 1 ns transition time and an amplitude slightly (~0.5 V) larger than the numerically calculated pull-in voltage as per the previously mentioned static analyses. Additionally, the pull-in time of the damped lumped-parameter model was correspondingly calculated by the MATLAB/Simulink model. Figure 8 plots the pull-in time as a function of the following: (a) the flexure width, w_x and w_y (stiffness); (b) the wall width w_w (mass); and (c) the normalized actuation voltage $V_o/V_{Pull-in}$. It should be noted that the analytical solutions based on Equation (8) also assumed $r = 0.5$ and calculated $V_{Pull-in}$ as per Equation (7); therefore, the discrepancy due to the approximate solution of the latter, as reported in the previous section, propagated to the plots of Figure 8.

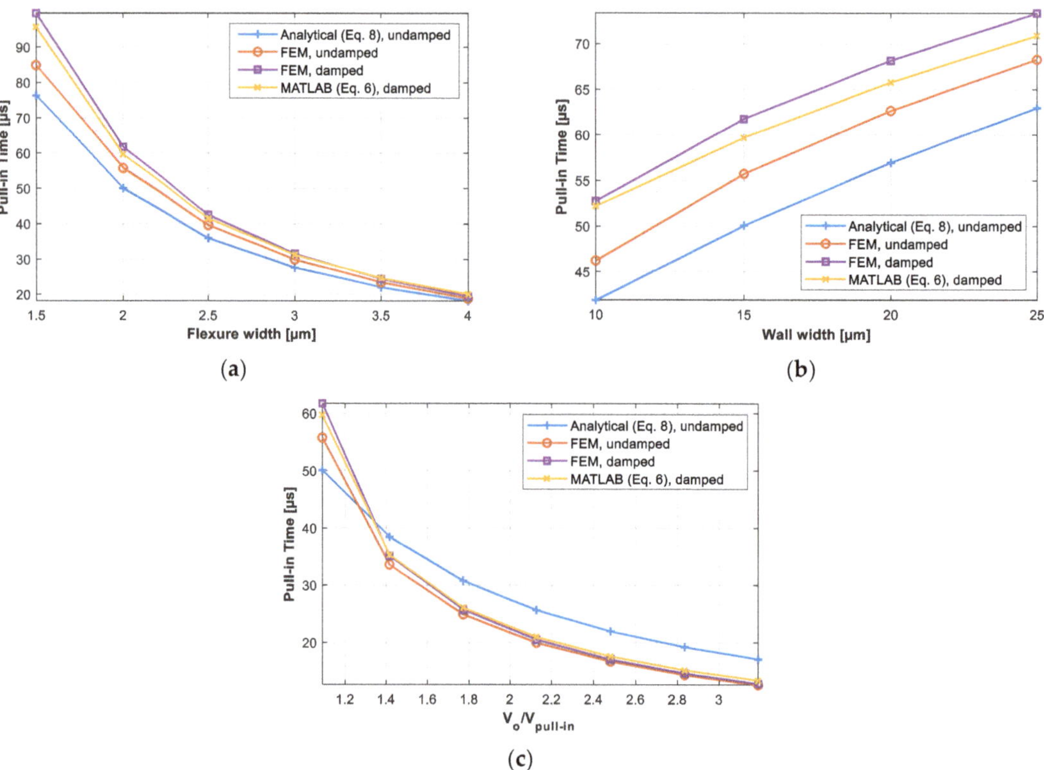

Figure 8. The results of transient parametric analyses of the pull-in time, where the latter is plotted as a function of the following: (**a**) the flexure width (by sweeping both w_x and w_y with the corresponding actuation voltage amplitude required for pull-in); (**b**) the wall width w_w; and (**c**) the normalized actuation voltage $V_0/V_{Pull-in}$ (the analytically calculated $V_{Pull-in} \cong 56.4$ V). In the plots, the 2D FEM results of the damped and undamped regimes are compared with the analytical solutions based on Equation (8). Additionally, the lumped-parameter model solutions by MATLAB/Simulink are also plotted.

As observed in Figure 8a and Figure 8b, Equation (8) and the FEM results correspond reasonably to one another. In the investigated ranges, the analytical solution consistently underestimates the pull-in time with differences between it and the corresponding undamped FEM simulation ranging roughly between 3.6% and 10.1% for the stiffness sweep (Figure 8a), generally decreasing with larger stiffnesses, and between 7.8% and 10.1% for the mass sweep (Figure 8b). Additionally, it should be pointed out that Equation (8) supposedly accounts for the time required to deflect the electrode along the full gap, whereas the FEM simulations had stoppers placed at 80% of the gap. Therefore, ideally, the numerically derived undamped curves should have remained below the analytical curves, which means that the error in the analytical account of the pull-in time is even larger than what the figures show. On the other hand, the results of the parametric sweep of the actuation voltage, shown in Figure 8c, reveal larger discrepancies with differences between 10.1% and (-) 35.8%, consistently increasing (absolute value) with larger actuation voltages.

Furthermore, the comparison between the FEM-based curves in Figure 8 shows that the effect of damping on the pull-in time is relatively small, and it becomes notably smaller when the applied actuation voltage is larger, regardless of the applicable pull-in voltage (Figure 8a,c). Additionally, by observing the curves obtained from the MATLAB model we find good quantitative agreements with the COMSOL curves, whereby absolute difference values ranging between 0.5% and 4% are found in the plotted ranges. The observed slight differences are expected as they result from the variations between the lumped parameters in the MATLAB model and the corresponding distributed parameters in the COMSOL model, given that the latter is naturally assumed to reflect the physical properties of the AUC more accurately.

However, a deeper examination of Equation (8) leads to the conclusion that the pull-in time is neutral to the stiffness element. This becomes evident by substituting Equation (7) and $\omega_0 = \sqrt{k_{eq}/m_{eq}}$ in Equation (8) and further simplifying the variables by using their basic constituents that correspond to the structure at hand, which provides the following:

$$t_{Pull\text{-}in} \cong \frac{2}{V_o} \sqrt{\frac{\rho d^3 A_{xy}}{\epsilon l_e}} , \tag{9}$$

where A_{xy} is the effective cross-sectional area of the movable electrode (excluding the anchor) in the xy-plane, that is $A_{xy} = A_{xy,wall} + r\,A_{xy,springs}$; ρ is the density of material; and l_e is the length of the stationary electrode, as defined in Figure 3 and Table 1. Therefore, it is inferred that the reduction in pull-in time with increased stiffness (flexure width), as seen in Figure 8a, in fact resulted from increasing the actuation voltage, which was needed to induce pull-in for larger flexure widths, rather than from increasing the stiffness.

On the other hand, contrary to the assumption made to derive Equation (8), the system at hand features viscous damping, which depends on the cube of the cell height (see Equation (5)). Accordingly, a parametric study was carried out with cell heights spreading from 10 μm to 100 μm (see Figure 9a). As expected, the cell height has no influence on the pull-in time of the undamped regime, since both the stiffness and mass of the structure, present in Equation (8), scale equally with height. This can be inferred from Equation (9), and it can also be observed by the identical undamped responses of the different cell heights in Figure 9a (dashed lines appearing as one). However, increasing the height showed a substantial change in pull-in time that increased more rapidly with larger heights, as observed in Figure 9b, which plots the normalized pull-in time as a function of the height. Here, the pull-in time is normalized relative to the pull-in time of the undamped structure (55.75 μs). It can be observed that the damping has negligible effect on the 10 μm high structure, with 0.4% increase in pull-in time (55.97 μs), but it has a substantial effect on the 100 μm high structure, with 58.6% increased pull-in time (88.41 μs) compared to the undamped regime. Moreover, solving the lumped-parameter model in MATLAB has confirmed the same correlation between the pull-in time and the cell height. As a result, it can be inferred that Equation (8) provides a good approximate evaluation of the pull-in time of the AUC only for small cell heights. Especially for large heights needed to achieve large forces for the cooperative system investigated here, the influence of damping has to be considered.

3.3 Pull-Out Time

For a cooperative system in which a large number of physically independent actuators have to be driven in a coordinated manner, it is not only the pull-in time of the individual actuators that has to be considered, but the so-called pull-out time, which is defined as the time needed to settle back from a pull-in situation to the unengaged starting position, must also be considered (see Figure 2). In order to estimate the pull-out time in FEM, first, a static study was carried out to bring the moveable electrode to the fully deflected position (being landed at the stoppers); then, the static solution was used as an initial condition for a transient study, which released the structure to oscillate freely under the influence of the structure's elasticity and the thin film's viscosity. In the literature, the pull-out time is usually defined as the time it takes the released electrode to first cross its undeflected position. This definition renders the stiffness and inertia components the major influencers of pull-out time, as they determine the oscillation frequency and, consequently, the recovering speed of the structure. However, the application at hand requires that the actuator be realigned between successive operations; hence, the settling time is of major consequence and regarding it as the measure of the pull-out time is more sensible.

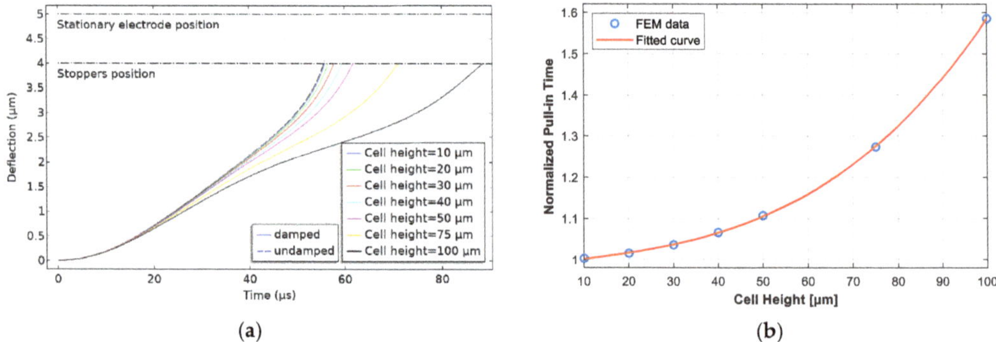

Figure 9. Transient parametric analysis of the pull-in time under both the damped and undamped regimes: (**a**) the deflection response of the AUC with a sweep of the cell height *h*; and (**b**) the corresponding pull-in time as a function of the cell height, where the time is normalized to the pull-in time of the undamped condition (55.75 μs).

Figure 10a plots a parametric study with a sweep of the flexure width. It shows how increasing the spring constant can substantially reduce the time required to first reach the undeflected position (dot-dashed line in the middle); in the investigated range, this time reduces from 46.4 μs (flexure width = 1.5 μm) down to 8.3 μs (flexure width = 4.0 μm). However, consideration of the decay rates of the curves' envelopes reveals equal settling times regardless of the structure's stiffness. On the other hand, a parametric study of the dimensional parameter, which has the strongest influence on the SQFD, i.e., the cell height, is shown in Figure 10b. The simulated cell heights, also ranging from 10 μm to 100 μm, show increasing tangible influence on the settling time of the damped oscillation motion as the cell height increases. The MATLAB/Simulink model also showed the same correlations of pull-in time with the flexure width and cell height.

In order to draw a sensible comparison between the curves in Figure 10b, their decaying rates were analyzed. Let us recall the well-known general solution for the response of an underdamped spring–mass–damper system, which takes the form $x(t) = ae^{-\zeta\omega_n t}\sin(\omega_d t + \varphi)$, where a and φ are the amplitude and phase of the response, and ζ, ω_n, and ω_d are the damping ratio, the natural frequency, and the damped natural frequency of the system, respectively. Accordingly, by fitting exponentially decaying envelopes of the form $ae^{-\zeta\omega_n t}$ to a certain point on these curves, with $a = 4\ \mu m$ and $\omega_n = 73,230\ rad/s$, "equivalent" damping ratios were extracted for comparison. The point of reference selected for this analysis was the point of tangency on the upper curvature of a response curve after a complete first oscillation, which experiences the highest dissipation rate compared to subsequent oscillations. Figure 10c demonstrates the analysis in which the response of the AUC with a cell height of 100 μm is plotted with the corresponding fitted decaying envelope, showing an equivalent $\zeta \cong 0.1466$. It should be noted that the aforementioned general solution corresponds to a constant damping coefficient that is linearly proportional to the velocity, which is not the case with the SQFD that depends additionally on the inverse of the cube of the changing thin film's width, as shown in the analytical modeling part (see Equation (5)). Therefore, it can be inferred that the decay rate in these oscillations is not constant but rather reduces as the deflection amplitude decreases with the continued oscillations. Moreover, Figure 10d shows the defined equivalent damping ratio as a function of the AUC's height where the resultant damping ratios are fitted with a line. Here, similarly to the response during pull-in, it can be inferred that the damping has a relatively negligible effect on the 10 μm high structure with $\zeta \cong 0.0061$. Additionally, the substantial influence of the cell height can also be observed by the several-orders-of-magnitude difference in the SQFD force as a consequence of changing this parameter, which is plotted in Figure 10e. Here, the largest SQFD force per unit length of the thin film (corresponds to l_e in Figure 3) exerted on the 10 μm high structure is about 1.0 (mN)/m in contrast to 348.6 (mN)/m experienced by the 100 μm high structure.

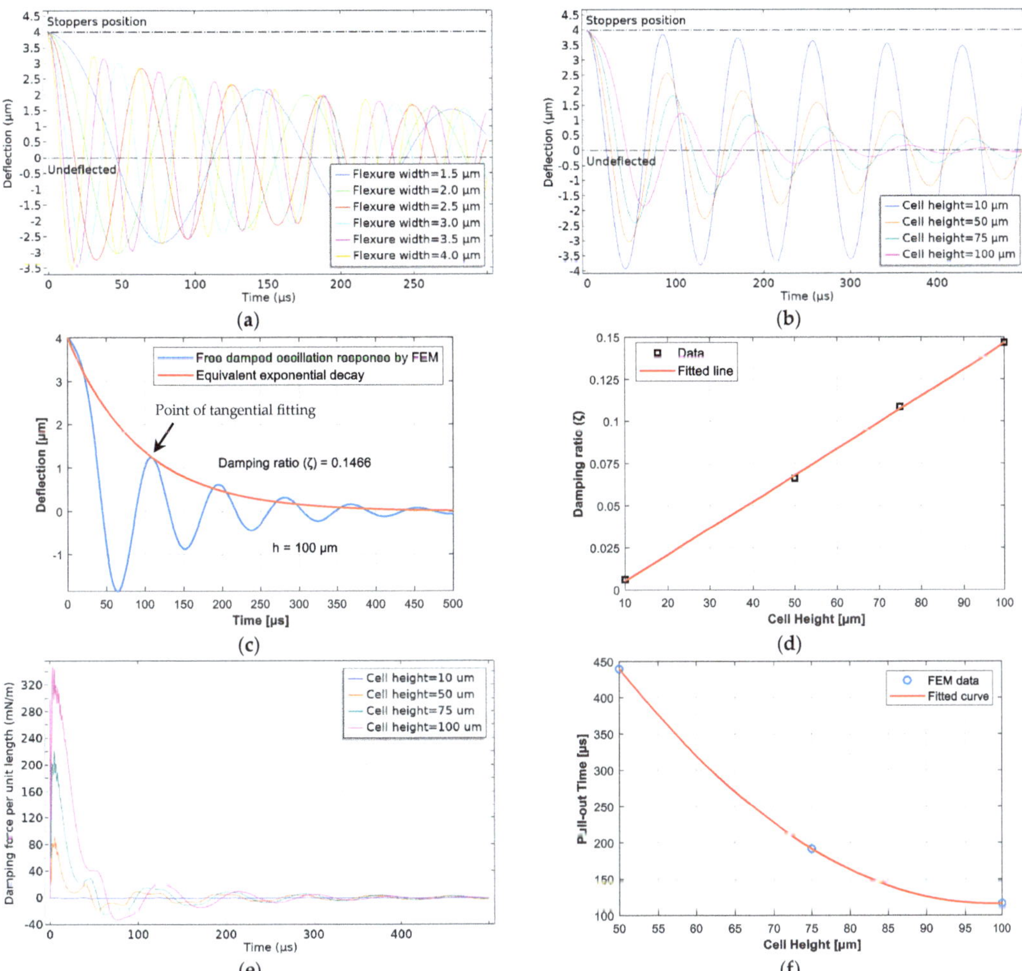

Figure 10. Transient parametric analyses of the pull-out time by the free damped oscillation response of the AUC after being released from the fully deflected position in a 2D FEM model corresponding to Table 1: (**a**) with flexures of various stiffnesses (via flexure width sweep, w_x and w_y) and (**b**) with a sweep of the cell height h. (**c**) An example of the analysis carried out on the responses shown in (**b**) to extract an equivalent damping ratio by fitting an exponentially decaying envelop to the upper curvature of the response after a complete first oscillation. The plotted response belongs to the 100 μm high structure. (**d**) The equivalent damping ratio due to SQFD as a function of the cell height. (**e**) The SQFD force per unit length of the thin-film exerted on the AUC with different cell heights. (**f**) An estimation of the pull-out time as a function of the cell height by applying a settling reference (a realignment margin δ) of ±1 μm to the response curves plotted in (**b**).

Additionally, by allowing a certain margin of deflection for the realignment of the AUC after an actuation cycle, below which the AUC is considered settled, the effect of the height on the dynamic response can be further assessed. For example, by setting a realignment margin (δ) of ±1 µm from the undeflected position, the settling time is plotted as a function of the height (see Figure 10f). It can be observed that the settling time drops more than threefold from 440 µs with the 50 µm high structure to 117 µs for the 100 µm high structure.

Furthermore, another possibility to increase SQFD that was explored here is by increasing the number of thin films damping the structure. This can be realized by fabricating additional damping walls adjacent to the frame of the moveable electrode from the inside, thereby the damping thin films can be effectively doubled. The FEM simulation carried out with a 50 µm high structure featuring additional inner damping thin films showed an increased damping ratio of $\zeta \cong 0.1027$ and a reduced settling time of 230 µs compared to $\zeta \cong 0.0662$ and 440 µs settling time without the additional films.

4. Discussion

By viewing the relatively brief history of inchworm motors, a number of characteristics and performance measures can be identified that newly proposed designs aim to improve, e.g., stability, position controllability, failure resistance (redundancy), delivered force, and displacement speed and range. Conceptually, the proposed system of cooperative actuators has good potential regarding all of the aforementioned aspects. For example, as observed, the shaft in this system can be actively actuated in two directions, allowing the exertion of force and the exact positioning control in both directions, which is not the case in many other inchworm motor designs that actively push a shaft in one direction and rely on a spring element to reset it to its initial position after some displacement. Nevertheless, achieving stable operation at a decent displacement speed for a system with many distributed cooperating actuators requires careful determination of the dynamic behavior of the system's basic actuator, the AUC.

4.1. Dynamic Behavior

By examining the pull-in time parametric analyses of the stiffness, inertia, and applied voltage, it can be concluded that the AUC does fulfill the inertia-limited assumption. The analyses demonstrated the dependencies described in Equation (8) with a decent qualitative agreement, with the exception of the applied voltage dependence, which Equation (8) reflected inaccurately as the difference percentages have shown. Additionally, with the moderate cell height used in those simulations, the damping showed relatively small effects. However, as revealed by the parametric analysis of the cell height, increasing this parameter can amplify the damping influence and increase the pull-in time by a considerable amount. On the other hand, compared with the pull-in time, the pull-out time analyses showed an even greater dependence on the damping, rendering the influence of the cell height even more critical. It is worth noting that, for instance, within the boundaries of the analyses shown in Figures 9 and 10, increasing the cell height from 50 µm to 100 µm increased the pull-in time by 26.6 µs (43%) but decreased the pull-out time by roughly 323 µs (73%). Hence, the sum of these transition times, which primarily defines the cycle time of the actuator, decreased from 501.8 µs to 205.4 µs, i.e., by 59%. Therefore, in contrast to the usual consideration of the pull-in time as the limiting parameter for the actuation speed of devices such as electrostatic cantilever microswitches [15] (p. 41), the pull-out time for the actuator at hand, which necessarily includes the settling time, is much longer than the pull-in time. Hence, reducing the pull-out time by larger damping at the expense of a slightly longer pull-in time ultimately reduces the cycle time of the AUC, allowing higher operating frequencies and faster displacement speeds for the cooperative system.

4.2. Output Force

The delivered forces by inchworm motors reported in the literature varied significantly. Yeh et al. [5] achieved 260 µN and Erismis et al. [7] achieved 110 µN at 33 V and 16 V driving voltages, respectively. On the other hand, Penskiy and Bergbreiter [8], who proposed the combination of the latching and driving mechanisms by a unidirectional movement of an inclined flexible arm, reported a force of 1.88 mN at 110 V. Moreover, in a recent publication, Teal et al. [10] claimed producing an impressive 15 mN of force at 100 V. In general, a common aspect among these actuators is the use of a GCA mechanism in a comb-drive arrangement that generally provides larger area for electrostatic force generation, whereas the proposed AUC structure has a single parallel-plate setup for the GCA instead, which requires some form of compensation in order to achieve the desired large force, e.g., higher structures or operating at smaller gaps between electrodes, which are limited by the available fabrication processes and the feasible aspect ratios. Additionally, the possibility of increasing the delivered force on the system level by implementing a larger number of AUCs is an advantage that the proposed concept facilitates, but this feature naturally has its own limits, e.g., overall size of the system and complexity of control and operation. Therefore, the amount of force delivered by the AUC has to be optimized within the limits of the available microfabrication technology, taking into account not only the design specifications of the cooperative system as a whole but also other tradeoffs on the level of the AUC itself.

For example, Figure 11 plots the electrostatic force exerted on the moveable electrode of an AUC corresponding to Table 1 when a voltage potential of 100 V is applied at Electrode 1 (see Figure 3). The figure also plots the corresponding counter elastic force by the springs. The net force output by the AUC, neglecting the transient damping effect, is the difference between the two curves. In the plot, three design parameters can be identified: the gap (distance to stationary electrode, $d = 5$ µm), the stroke (distance to stoppers, $s = 4$ µm), and the distance at which the interlocking teeth of the shaft is first engaged by the corresponding teeth of the AUC, which equals 1 µm (see Figure 1). If one defines the force of the AUC by the minimum force delivered to the shaft during its displacement, which is the force at the first moment of engagement, then the selected parameters result in a force of 0.046 mN. Note that the effective stroke that the AUC moves the shaft by in this arrangement is 3 µm. On the other hand, if the AUC engages the shaft at 3 µm distance from the undeflected position, allowing an effective stroke of 1 µm, the force output increases to 0.202 mN. From this example analysis, it is inferred that on the AUC's level, there is a clear tradeoff between the force and the effective stroke, which, in turn, translates to another tradeoff on the system level between the number of AUCs required to achieve a certain output force and the number of cycles required to achieve a certain shaft displacement, which also affects the required frequency of operation to achieve a certain displacement speed. Therefore, an optimization study for the force of the AUC considering the aforementioned aspects as well as other important considerations, e.g., the stability of operation, is planned for the future.

4.3. Fabrication Feasibility Analysis

As discussed, the dynamic behavior (defined mainly by the pull-in and pull-out times) and the force provided by a single AUC strongly depend on geometrical parameters and, thus, on the limitations of the fabrication process. A typical set of the most critical parameters, which allow a reasonable compromise between large force, high speed, and reliability of the system, is shown in Table 2. Obviously, such geometries belong to High Aspect Ratio Microsystems (HARMS) and need appropriate processes providing the anisotropy, e.g., of etching. Considering the system approach shown in Figure 1, a SOI process is chosen for this feasibility study. With SOI, all structures shown in Figure 1 will be made out of crystalline silicon. In this case, the critical structure height is defined by the device layer thickness, the handle layer thickness, or the sum of both, and the critical lateral dimensions (gap, interlocking teeth, stoppers, etc.) are defined by lithography and Deep

Reactive Ion Etching (DRIE), e.g., by the Bosch process [24]. The required free-standing structures (suspensions and movable electrodes) can be provided by a combination of Surface Micromachining (SMM) using buried oxide (BOX) as the sacrificial layer and a DRIE step from the front side or from the back side for the areas which should be released on a large area.

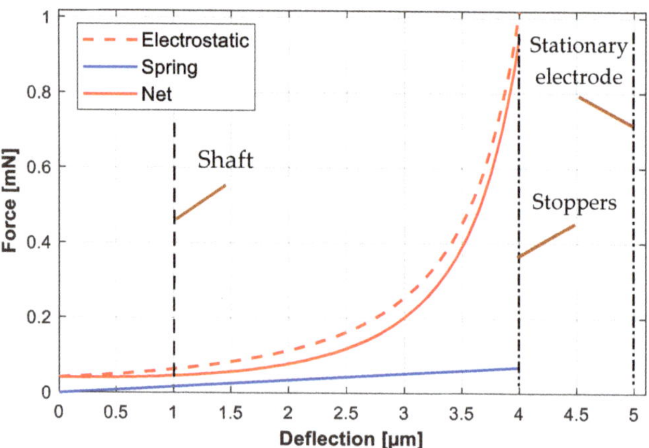

Figure 11. The electrostatic, elastic and net forces developed in an AUC corresponding to the parameters listed in Table 1 when subjected to an actuation voltage amplitude of 100 V.

The feasibility of such processes was demonstrated by one of the authors in the realization of a 3D-energy harvester [25] and an electrostatically driven bistable actuator for large forces and large stroke using the toggle-level principle [26]. Whereby, in the former, a thickness of a seismic mass of 411 µm was obtained by performing DRIE from both sides of a SOI-wafer (handle layer thickness of 400 µm, BOX thickness of 1 µm, and device layer thickness of 10 µm), and a minimum gap of 9.5 µm was defined by standard proximity printing in combination with DRIE. In the latter, a device layer thickness of 10 µm and suspensions width and length of 2 µm and 700 µm, respectively, were realized, similarly to the serpentine springs proposed for the AUC. However, fabricating a much larger device layer thickness and, accordingly, height of suspensions, electrodes, etc., is also possible, as demonstrated in [27] where a symmetric SOI wafer with an equal thickness of 100 µm for the device and handle layers and a BOX thickness of 2 µm in between was used, thus, providing structural heights of 100 µm and 202 µm by performing DRIE either on the device layer only or on both the device and handle layers.

With respect to the critical dimensions and the smallest gaps between electrodes and neighboring structures, e.g., stoppers, the resolution and the overlay accuracy of the used lithography process has to be considered for the feasibility study. Standard Deep Ultraviolet (DUV) lithography is suitable for the presented geometries and meets the accordingly needed overlay accuracy, e.g., for a typical DUV stepper a resolution of 0.25 µm and an overlay accuracy of <45 nm is specified [28]. A DUV stepper provides a relatively small field size of about 20 mm^2, which will limit the total system size and, thus, the possible number of AUCs that can be integrated in the system to provide large forces, if stitching several exposure fields for structuring the entire actuator system is avoided. However, on an experimental basis, direct laser writing can be used. A system available in our technology provides a resolution of about 0.6 µm with linewidth control of 80 nm and alignment accuracy of 200 nm (both 3σ values) over a field size up to 200 × 200 mm^2. These values meet the lithography requirements for defining the geometries presented in Table 2.

Table 2. Examples of proposed AUC geometries for fabrication and their estimated performance characteristics. Absent AUC parameters have the same values as those listed in Table 1.

Description	Symbol	Device 1	Device 2	Unit
Height of unit cell	h	50	100	µm
Width of cell wall	w_w	25	35	µm
Size of cell wall (inner)	c_x and c_y	500	600	µm
Size of anchor	a_x and a_y	100	100	µm
Width of spring (flexure width)	w_x and w_y	2.5	5	µm
Length of connector beams	$l_{c,x}$ and $l_{c,y}$	50	50	µm
Length of span beams	$l_{s,x}$ and $l_{s,y}$	200	300	µm
Length of extension beams	$l_{e,x}$ and $l_{e,y}$	53	82.5	µm
Length of stationary electrodes	l_e	480	600	µm
Width of stopper	w_s	25	25	µm
Nominal air gap between electrodes	d	5	4	µm
Stroke	s	4	3	µm
Equivalent mass (assuming $r = 0.5$)	m_{eq}	6.46	22.65	µg
Equivalent spring constant [1]	k_{eq}	40.4	304.8	N/m
Pull-in voltage [1]	$V_{pull\text{-}in}$	83.9	104.3	V
Collapse voltage (2nd pull-in) [1]	$V_{collapse}$	152.0	165.9	V
Actuation voltage	V_0	110	130	V
Pull-in time (corresponding to V_0) [2]	$t_{pull\text{-}in}$	32.2	24.9	µs
Net force of AUC (at 1 µm) [2,3]	F	0.04	0.19	mN
Net force of AUC (at 2 µm) [2,3]	F	0.06	0.51	mN
Realignment margin	δ	±2	±1	µm
Pull-out time (corresponding to δ) [2]	$t_{pull\text{-}out}$	283	91	µs

[1] Estimation based on static analysis in COMSOL 2D model. [2] Rough estimation based on MATLAB lumped-parameter model. [3] Damping force is neglected; the distance is from the undeflected position to the shaft (Figure 11).

However, to achieve large structure heights of 100 µm in combination with small structure widths in the presence of very small gaps between structures (<5 µm), the DRIE process conditions have to be optimized with respect to the specific layout (loading effect) and the needed large anisotropy. The DRIE machine at our disposal is capable of producing structures with aspect ratios of 1:20 with a minimum feature size of 2 µm. These limitations were taken into consideration for the geometries shown in Table 2. It is worth pointing out that the stoppers and stationary electrodes of the AUC will have larger dimensions in the outward directions than what is shown in Figure 3 (the simulation model) such that enough contact area with the BOX layer is realized to ensure stability, especially after the latter is etched to free the moveable electrode. Additionally, slight deviations from vertical slope angles of sidewalls have to be considered in modelling such systems. Therefore, the presented COMSOL model has to be extended in the future to a full 3D model.

4.4. Error Estimation for the Used SQFD Modeling Approach in COMSOL

As previously mentioned, in order to obtain numerical models that more easily converge and that consume less time for more rapid simulations, the primary FEM modeling used in this paper was 2D. Moreover, as mentioned in Section 2.3.2., modeling the SQFD via the COMSOL built-in BC in the 2D environment was not possible; hence, the derived analytical solution, i.e., Equation (5), was embedded in the model as a boundary load. To investigate the error introduced by these simplifications, comparisons between the two approaches were made. To this end, the deflection responses of the AUC under two actuation voltage amplitudes were simulated in a 3D COMSOL model. Figure 12a plots the deflection responses under an applied voltage that is enough for inducing pull-in for two structures with different heights, along with the undamped response. The built-in BC "Thin-Film Damping" is solved for with the modified Reynolds equation, where the default settings for the fluid-film properties and border nodes are applied. For the 50 µm high structure, the pull-in times calculated for the damped responses of the embedded analytical model and the built-in BC are 66.7 µs and 66.6 µs, respectively, whereas they are 90.0 µs and 88.7 µs in the same order for the 100 µm high structure. These results show very good correlations with less than 1.5% maximum discrepancy in the case of the higher structure. Moreover, Figure 12b plots the damped deflection responses according to the two approaches under an applied voltage of 55 V, which is less than the pull-in voltage (~61 V). The plot shows near identical responses. As a result, we conclude that embedding the analytical model of the SQFD described in Section 2.2. In the 2D FEM simulations is a decent substitute to the appropriate COMSOL built-in BC; hence, the simulation results obtained by this approach, as presented in chapter 3, describe the system behavior adequately.

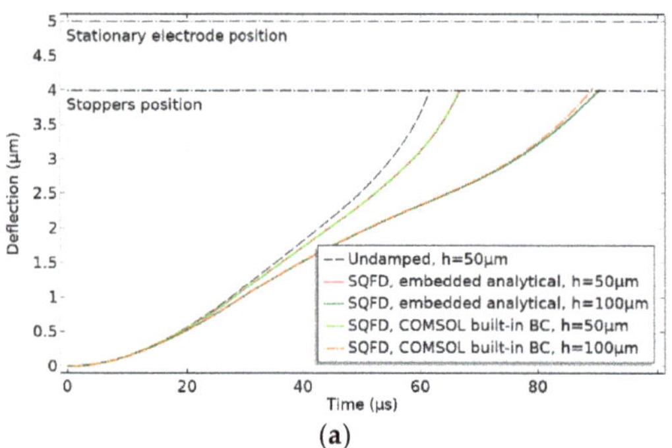

(a)

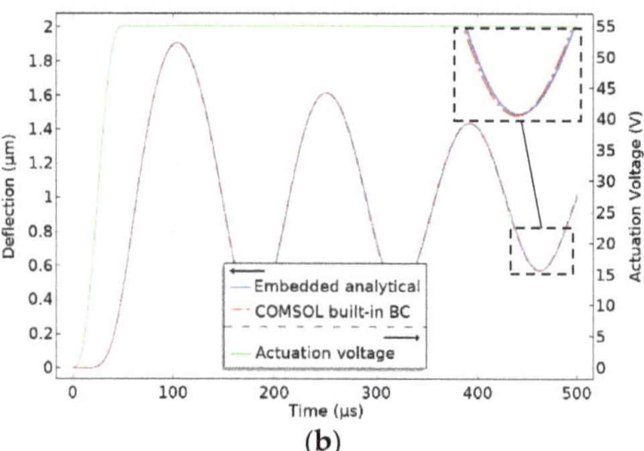

(b)

Figure 12. Comparison of the FEM simulations for the SQFD effect on a 3D model of the AUC in COMSOL. The SQFD applied via the COMSOL built-in BC is shown in contrast to the embedded analytical solution presented in this paper, which is implemented as a boundary load. (**a**) The damped deflection responses to an applied step voltage with $V_0 = 61.5$ V and a transition time of 1 ns for two structures with different heights, $h = 50$ µm and 100 µm, along with the undamped response (essentially independent of h). (**b**) The damped deflection responses to an applied step voltage of $V_0 = 55$ V and a transition time of 50 µs for a cell height $h = 50$ µm. In these simulations, the wall thickness $w_w = 20$ µm, whereas the other dimensions of the AUC are as listed in Table 1. Here, the pull-in voltage is ≈61 V.

5. Conclusions

In this work, a concept for a modular-like cooperative actuator system that is inspired by the inchworm motor is presented. Moreover, an analytical model for the AUC, which is the proposed actuator for the cooperative system, is described. Furthermore, the results of couple-field numerical modeling by COMSOL for the behavior of the AUC are presented, and comparisons with analytical models from the literature are drawn where possible. The results of the static FEM analyses carried out on the AUC showed excellent correlation with the analytical model for the pull-in voltage found in the literature. On the other hand, the transient analyses of the pull-in time showed decent qualitative correlations with a model of an inertia-limited system found in the literature; however, significant discrepancies were observed when the damping in the system is amplified by structures of larger heights. Additionally, by comparison with the FEM results, it was concluded that the dependence of the pull-in time on the applied voltage is not adequately represented by the referenced model. The results of the transient numerical parametric analyses for the pull-in and pull-out times were validated by solving the corresponding lumped-parameter model in MATLAB/Simulink.

Moreover, the transient analyses of the pull-in and pull-out times, which constitute the major parts of the AUC's actuation cycle, revealed substantial dependence on the SQFD, which, in turn, is greatly influenced by the cell height. It was found that the dynamic behavior of the proposed system is most critically influenced by the pull-out time that is relatively much longer than the pull-in time. An example analysis showed that by doubling the cell height (from 50 µm to 100 µm), the damping profile in the actuator increased considerably, and the pull-out time decreased by more than tenfold the amount of time the pull-in time increased. Additionally, a tradeoff between produced forces and speed of the inchworm actuation system was found. Although a single AUC is expected to provide forces in the range of 0.1 mN to 0.5 mN, for a cooperative system with many AUCs working in parallel, forces of several tens of millinewtons can be expected when typical design rules for SOI-technology using standard photolithography and DRIE are considered. As a result, some guidelines for the design of the AUC can be drawn from the presented study such that the major performance characteristics, e.g., cycle time, frequency of operation, and force of the actuator, can be defined. In the future, the AUC is intended to be fabricated, and experimental-based analyses of the parameters investigated here will be carried out to establish the validity of the models and simulations presented in this paper.

Author Contributions: Conceptualization, A.A. and U.M.; methodology, A.A. and U.M.; software, A.A.; validation, A.A.; formal analysis, A.A.; investigation, A.A.; resources, U.M.; data curation, A.A.; writing—original draft preparation, A.A.; writing—review and editing, A.A. and U.M.; visualization, A.A.; supervision, U.M.; project administration, U.M.; funding acquisition, U.M. All authors have read and agreed to the published version of the manuscript.

Funding: This research is funded by the German Research Foundation (Deutsche Forschungsgemeinschaft—DFG) under the umbrella of the priority program SPP2206—Cooperative Multistage Multistable Micro Actuator Systems (KOMMMA), project number ME 2093/5-1. Additionally, this open-access publication was partially funded by the Ministry of Science, Research and Arts in Baden-Württemberg, Germany.

Institutional Review Board Statement: Not applicable.

Informed Consent Statement: Not applicable.

Data Availability Statement: Not applicable.

Acknowledgments: The authors are grateful for the conceptualization work of the actuator system accomplished by Hussam Kloub for the aforementioned research project, as shown in Figure 1.

Conflicts of Interest: The authors declare no conflict of interest. Moreover, the funders had no role in the design of the study; in the collection, analyses, or interpretation of data; in the writing of the manuscript; or in the decision to publish the results.

Appendix A

Similar to the approach by Fedder [29] (pp. 84–110), the analytical formulations for the longitudinal and lateral spring constants of the serpentine flexures utilized in the AUC were derived from free body diagrams by using Castigliano's second theorem. The derivations were made by considering only bending forces within the linear elastic regime and the Euler–Bernoulli beam model [30] (pp. 14–23).

Accordingly, the longitudinal spring constant of a serpentine flexure, such as the one shown in Figure A1a, equals the following:

$$k_{lg} = \frac{E I_z}{\lambda}, \tag{A1}$$

such that

$$I_z = \frac{h w^3}{12}, \tag{A2}$$

and

$$\lambda = \tfrac{8}{3}\beta^2 \left(l_e{}^3 + l_c{}^3\right) + \tfrac{1}{6}l_s{}^3 + 2 l_e{}^2 \beta^2 \left(4 l_c + l_s - 2\varphi\right)$$
$$+ l_c{}^2 \beta^2 \left(8 l_e + l_s \left(3 - \tfrac{1}{\beta}\right) - 4\varphi\right) + \tfrac{1}{2} l_s{}^2 l_c \left(1 - \beta\right) \tag{A3}$$
$$+ 2\varphi^2 \beta^2 \left(l_e + l_c + l_s\right) + 4\beta^2 \left(l_e l_c l_s - \varphi \left(2 l_e l_c + l_e l_s + l_c l_s\right)\right),$$

where

$$\varphi = \frac{l_e{}^2 + l_c{}^2 + 2 l_e l_c + l_e l_s + l_c l_s}{l_e + l_c + l_s} \tag{A4}$$

and

$$\beta = \frac{6l_c{}^3l_s + 3l_cl_s{}^3 + 9l_c{}^2l_s{}^2 + 6l_el_c{}^2l_s + 3l_el_cl_s{}^2}{\left(2l_e{}^4 + 2l_c{}^4 + 8l_e{}^3l_c + 2l_e{}^3l_s + 8l_el_c{}^3 + 5l_c{}^3l_s + 12l_e{}^2l_c{}^2 + 6l_e{}^2l_cl_s + 9l_el_c{}^2l_s + 3l_c{}^2l_s{}^2\right)}.\tag{A5}$$

On the other hand, the lateral spring constant equals the following:

$$k_{lt} = \frac{EI_z}{\gamma},\tag{A6}$$

such that

$$\begin{aligned}\gamma = {} &\tfrac{8}{3}\left(l_e{}^3 + l_c{}^3\right) + \tfrac{1}{6}l_s{}^3\alpha^2 + 2l_e{}^2\left(4l_c + l_s - 2\varphi\right)\\ &+l_c{}^2\left(8l_e + l_s\left(3 - \alpha\right) - 4\varphi\right) + \tfrac{1}{2}l_s{}^2l_c\alpha\left(\alpha - 1\right)\\ &+2\varphi^2\left(l_e + l_c + l_s\right) + 4\left(l_el_cl_s - \varphi\left(2l_el_c + l_el_s + l_cl_s\right)\right),\end{aligned}\tag{A7}$$

and

$$\alpha = \frac{6l_c{}^2 + 3l_cl_s}{2l_s{}^2 + 6l_cl_s},\tag{A8}$$

where l_e, l_c, and l_s are the lengths of the extension, connector, and span beams, respectively; and w is the serpentine flexure width (w_x or w_y) (see Table 1 and Figure 3).

Consequently, the four parallel-connected suspension springs that constitute the stiffness element of the AUC, assuming the wall is rigid, possess an equivalent spring constant along the x-axis that can be written as follows:

$$k_{eq} = 2\,k_{x,lg} + 2\,k_{y,lt},\tag{A9}$$

where the subscripts x and y denote the axis along which the serpentine flexure extends, and the applicable longitudinal or lateral spring constants are denoted by lg or lt, respectively. It follows that to obtain the equivalent spring constant in the y-axis, the axes' subscripts on the right-hand side are switched.

Moreover, in Figure A1b, an example of the validation of the analytically derived formulations against a 3D FEM model is shown. Here, the length of the span beam of the serpentine flexure is swept between 20 µm and 200 µm. It is worth noting that taking the bending forces solely into account under the assumption of the Euler–Bernoulli beam model is considered valid for long beams where the length is at least 5–7 times larger than the largest of the other two dimensions. Therefore, the model becomes less accurate with shorter beams (larger discrepancy between the blue curves in the left hand side of Figure A1b) and higher serpentine structures, in which case it becomes more important to take the shearing deformations into account according to Timoshenko's beam model [30] (p. 17).

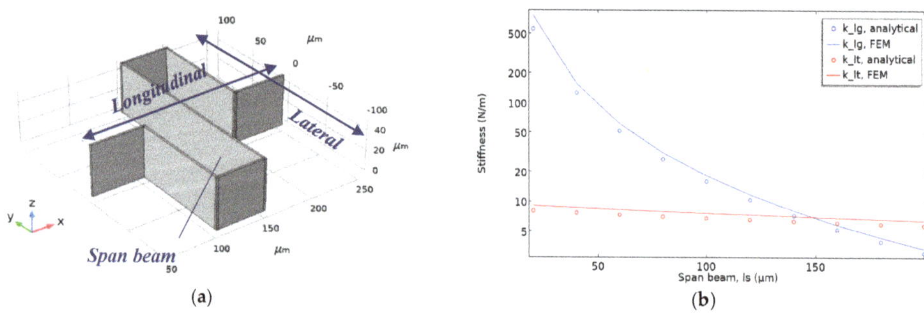

Figure A1. (**a**) A 3D model of the serpentine flexure design used in the AUC with dimensions corresponding to Table 1, plotted in COMSOL. (**b**) A comparison of the analytically derived and 3D-FEM-based longitudinal and lateral spring constants with a sweep of the span beam of the serpentine flexure shown in (**a**). The longitudinal spring constant shows greater sensitivity to the span length than the lateral spring constant does.

References

1. Neul, R.; Gmez, U.-M.; Kehr, K.; Bauer, W.; Classen, J.; Dring, C.; Esch, E.; Gtz, S.; Hauer, J.; Kuhlmann, B.; et al. Micromachined Angular Rate Sensors for Automotive Applications. *IEEE Sensors J.* **2007**, *7*, 302–309. [CrossRef]
2. Sampsell, J.B. Digital micromirror device and its application to projection displays. *J. Vac. Sci. Technol. B* **1994**, *12*, 3242. [CrossRef]
3. Zhang, W.-M.; Meng, G.; Chen, D.I. Stability, Nonlinearity and Reliability of Electrostatically Actuated MEMS Devices. *Sensors* **2007**, *7*, 760–796. [CrossRef]
4. Dochshanov, A.; Verotti, M.; Belfiore, N.P. A Comprehensive Survey on Microgrippers Design: Operational Strategy. *J. Mech. Des.* **2017**, *139*, 271. [CrossRef]
5. Yeh, R.; Hollar, S.; Pister, K. Single mask, large force, and large displacement electrostatic linear inchworm motors. *J. Microelectromech. Syst.* **2002**, *11*, 330–336. [CrossRef]
6. Kim, S.-H.; Hwang, I.-H.; Jo, K.-W.; Yoon, E.-S.; Lee, J.-H. High-resolution inchworm linear motor based on electrostatic twisting microactuators. *J. Micromech. Microeng.* **2005**, *15*, 1674–1682. [CrossRef]
7. Erismis, M.A.; Neves, H.P.; Puers, R.; van Hoof, C. A Low-Voltage Large-Displacement Large-Force Inchworm Actuator. *J. Microelectromech. Syst.* **2008**, *17*, 1294–1301. [CrossRef]
8. Penskiy, I.; Bergbreiter, S. Optimized electrostatic inchworm motors using a flexible driving arm. *J. Micromech. Microeng.* **2013**, *23*, 15018. [CrossRef]
9. Contreras, D.S. Walking Silicon: Actuators and Legs for Small-Scale Terrestrial Robots. Ph.D. Dissertation, University of California, Berkeley, CA, USA, 2019.
10. Teal, D.; Gomez, H.C.; Schindler, C.B.; Pister, K.S.J. Robust Electrostatic Inchworm Motors for Macroscopic Manipulation and Movement. In Proceedings of the 2021 21st International Conference on Solid-State Sensors, Actuators and Microsystems (Transducers), Orlando, FL, USA, 20–24 June 2021; pp. 635–638. [CrossRef]
11. Xiang, X.; Dai, X.; Wang, K.; Yang, Z.; Sun, Y.; Ding, G. A Customized Nonlinear Micro-Flexure for Extending the Stable Travel Range of MEMS Electrostatic Actuator. *J. Microelectromech. Syst.* **2019**, *28*, 199–208. [CrossRef]
12. Burns, D.M.; Bright, V.M. Nonlinear flexures for stable deflection of an electrostatically actuated micromirror. In *Microelectronic Structures and MEMS for Optical Processing III, Proceedings of the Micromachining and Microfabrication, Austin, TX, USA, 29 September 1997*; Motamedi, M.E., Herzig, H.P., Eds.; SPIE: Bellingham, WA, USA, 1997; Volume 3226, pp. 125–136. [CrossRef]
13. Zhang, W.-M.; Yan, H.; Peng, Z.-K.; Meng, G. Electrostatic pull-in instability in MEMS/NEMS: A review. *Sens. Actuators A Phys.* **2014**, *214*, 187–218. [CrossRef]
14. Younis, M.I. *MEMS Linear and Nonlinear Statics and Dynamics*; Springer US: Boston, MA, USA, 2011; ISBN 978-1-4419-6019-1.
15. Ostasevicius, V.; Dauksevicius, R. *Microsystems Dynamics*; Springer: Dordrecht, The Netherlands, 2011; ISBN 978-90-481-9700-2.
16. Kloub, H. Design Concepts of Multistage Multistable Cooperative Electrostatic Actuation System with Scalable Stroke and Large Force Capability. In Proceedings of the ACTUATOR; International Conference and Exhibition on New Actuator Systems and Applications 2021, Online, 17–19 February 2021; pp. 1–4.
17. Teli, M.; Grava, G.; Solomon, V.; Andreoletti, G.; Grismondi, E.; Meswania, J. Measurement of forces generated during distraction of growing-rods in early onset scoliosis. *World J. Orthop.* **2012**, *3*, 15–19. [CrossRef] [PubMed]
18. Bao, M.; Yang, H. Squeeze film air damping in MEMS. *Sens. Actuators A Phys.* **2007**, *136*, 3–27. [CrossRef]
19. Andrews, M.; Harris, I.; Turner, G. A comparison of squeeze-film theory with measurements on a microstructure. *Sens. Actuators A Phys.* **1993**, *36*, 79–87. [CrossRef]
20. Blech, J.J. On Isothermal Squeeze Films. *J. Lubr. Technol.* **1983**, *105*, 615–620. [CrossRef]
21. Starr, J.B. Squeeze-film damping in solid-state accelerometers. In Proceedings of the IEEE 4th Technical Digest on Solid-State Sensor and Actuator Workshop, Hilton Head, SC, USA, 4–7 June 1990; pp. 44–47.
22. Chu, P.B.; Nelson, P.R.; Tachiki, M.L.; Pister, K.S. Dynamics of polysilicon parallel-plate electrostatic actuators. *Sens. Actuators A Phys.* **1996**, *52*, 216–220. [CrossRef]
23. Rebeiz, G.M. *RF MEMS: Theory, Design, and Technology*; Wiley InterScience: Hoboken, NJ, USA, 2003; ISBN 0-471-20169-3.
24. Laermer, F.; Schilp, A. Method of Anisotropically Etching Silicon. U.S. Patent 5,501,893, 27 November 1993.
25. Mescheder, U.; Nimo, A.; Müller, B.; Elkeir, A.S.A. Micro harvester using isotropic charging of electrets deposited on vertical sidewalls for conversion of 3D vibrational energy. *Microsyst. Technol.* **2012**, *18*, 931–943. [CrossRef]
26. Freudenreich, M.; Mescheder, U.; Somogyi, G. Simulation and realization of a novel micromechanical bi-stable switch. *Sens. Actuators A Phys.* **2004**, *114*, 451–459. [CrossRef]
27. Kronast, W.; Mescheder, U.; Müller, B.; Nimo, A.; Braxmaier, C.; Schuldt, T. Development of a tilt actuated micromirror for applications in laser interferometry. In *MOEMS and Miniaturized Systems IX, Proceedings of the MOEMS MEMS, San Francisco, CA, USA, 23 January 2010*; Schenk, H., Piyawattanametha, W., Eds.; SPIE: Bellingham, WA, USA, 2010; Volume 7594, p. 75940O. [CrossRef]
28. De Zwart, G.; van den Brink, M.A.; George, R.A.; Satriasaputra, D.; Baselmans, J.; Butler, H.; van Schoot, J.B.; de Klerk, J. Performance of a step-and-scan system for DUV lithography. In *Optical Microlithography X. Proceedings of Microlithography '97, Santa Clara, CA, USA, 10 March 1997*; Fuller, G.E., Ed.; SPIE: Bellingham, WA, USA, 1997; Volume 3051, p. 817. [CrossRef]
29. Fedder, G.K. Simulation of Microelectromechanical Systems. Ph.D. Dissertation, University of California, Berkeley, CA, USA, 1994.
30. Lobontiu, N.; Garcia, E. *Mechanics of Microelectromechanical Systems*; Kluwer Academic: Boston, MA, USA, 2005; ISBN 0-387-23037-8.

Article

Large Stepwise Discrete Microsystem Displacements Based on Electrostatic Bending Plate Actuation

Lisa Schmitt * and Martin Hoffmann *

Microsystems Technology (MST), Faculty of Electrical Engineering and Information Technology, Ruhr-University Bochum (RUB), 44801 Bochum, Germany
* Correspondence: lisa.schmitt-mst@rub.de (L.S.); martin.hoffmann-mst@rub.de (M.H.)

Abstract: We present the design, fabrication, and experimental characterization of microsystems achieving large and stepwise discrete displacements. The systems consist of electrostatic bending plate actuators linked in a chain with increasing electrode gaps to allow a stepwise system displacement. A derived analytic transfer function permits to evaluate the influence of the system components on both the total and the stepwise system displacement. Based on calculation and simulation results, systems featuring 5, 8, 10, 13, and 16 steps are modeled and fabricated using a dicing-free SOI-fabrication process. During experimental voltage- and time-dependent system characterization, the minimum switching speed of the electrostatic actuators is 1 ms. Based on the guiding spring stiffness and the switching time, step-by step and collective activations of the microsystems are performed and the system properties are derived. Furthermore, we analyze the influence of the number of steps on the total system displacement and present 16-step systems with a total maximum displacement of 230.7 $\pm$ 0.9 µm at 54 V.

Keywords: large stepwise discrete displacements; bending plate actuators; dicing-free SOI-technology; MEMS

Citation: Schmitt, L.; Hoffmann, M. Large Stepwise Discrete Microsystem Displacements Based on Electrostatic Bending Plate Actuation. *Actuators* **2021**, *10*, 272. https://doi.org/10.3390/act10100272

Academic Editors: Manfred Kohl, Stefan Seelecke and Ulrike Wallrabe

Received: 9 September 2021
Accepted: 12 October 2021
Published: 15 October 2021

1. Introduction

The presented multistage and multistable actuator system with a large stroke was motivated by terahertz (THz) applications. For THz systems, a true time delay is required, usually subdivided in binary fractions of the wavelength [1]. To achieve such a true time delay of THz beam steering arrays, a stepwise high-throw actuator microelectromechanical system (MEMS) would be a perfect solution.

Due to the fast response to electric signals [2] and the suitability for harsh environments [3,4], electrostatic actuators are used in THz systems [5] for optical switching [6], capacitors [7], micromirrors [8], and biomedical applications [9]. The usage in underwater environments [3] and in wearable haptics [10] enlarges the application field and increases the demand for large displacement and quick response electrostatic actuators.

Electrostatic actuators appear in many different varieties, of which the comb-drive and the parallel-plate actuators are the best known. Legtenberg et al. [11], amongst others, were pioneers showing the importance of the guiding springs enlarging the displacement of the comb-drives. By adapting the shuttle geometry, Grade et al. [2] demonstrated that these actuators can achieve displacements up to 150 µm in less than 1 ms with operating voltages below 150 V.

Electrostatic parallel-plate actuators reduce the distance between the electrodes in response to an applied voltage [12]. Various investigations [13–17] to reach a stable and large displacement by controller architectures resulted in a high input voltage and a sensitivity to parasitic capacitance. Thus, by adding an isolating layer between the electrodes and by adapting the electrode design, large and stable displacements can be achieved. Based on this, in [6], deflections exceeding 62.5 µm are reported at a driving voltage below 30 V. Therefore, a very long flexible electrode was combined with a rigid cantilever.

Preetham et al. [3] report a peak-to-peak displacement of 19.5 μm at ±8 V using curved isolated electrodes. In [18], we present a 3-bit digital-to-analog converter achieving a displacement of 149.5 μm at 60 V. In [19], we present three bending plate actuators linked in a chain and achieving a stepwise displacement of up to 35.8 μm at 50 V.

In this contribution, we present microsystems consisting of numerous flexible electrostatic bending plate actuators linked in a chain and achieving large displacements of up to 230.7 ± 0.9 μm at 54 V. In Section 2, we present the system function and derive the analytical transfer function to model microsystems with multiple steps. We focus on the sinusoidal guiding springs [20] and their huge influence on the system displacement. In Section 3, we show the fabrication of the systems featuring 5, 8, 10, 13, and 16 steps which are based on a dicing-free SOI-technology [21–23]. In Section 4, we characterize the bending plate actuators with a focus on the influence of the electrode geometry. Then, step-by-step and collective actuator activations as well as the switching time are presented. Finally, the system behavior depending on the number of steps and the guiding spring is analyzed. The main achievements of this work are summarized in Section 5.

2. Concept

2.1. Flexible Electrostatic Bending Plate Actuators

The required displacement of about 20 μm is generated by flexible electrostatic bending plate actuators. These actuators are based on a rotor electrode (RE) and a stator electrode (SE) facing each other at the distance $x_{electrode}$ (Figure 1a). The rotor electrode is fixed to a spring-mounted rotor. The stator electrode is combined with a flexible cantilever. When a voltage is applied between the electrodes, the tips of the electrodes approach as soon as the electrostatic force between the electrodes exceeds the overall mechanical force of springs and electrodes. Therefore, the electrodes bend towards each other which is the origin to name these actuators bending plate actuators. The bending of the electrodes starts with the tips approaching each other as the electrodes feature the smallest stiffness at the tips. This behavior is examined experimentally in Section 4.2.2. With increasing voltage, the cantilever of the stator electrode bends down the distance b. The pull-in is completed as soon as the electrodes are completely in contact, as shown in Figure 1b.

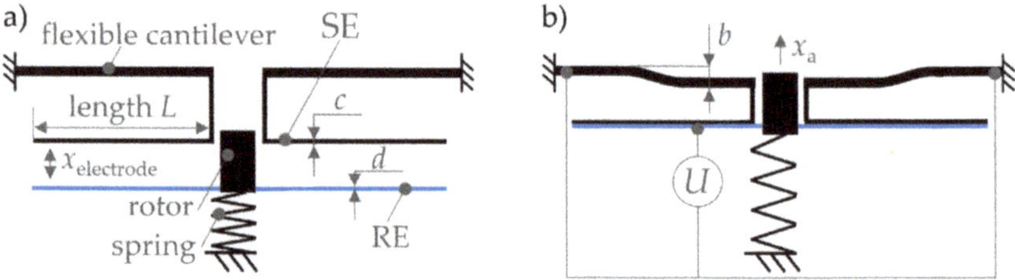

Figure 1. (**a**) Setup of the bending plate actuator, (**b**) completely pulled-in bending plate actuator, (drawings not to scale).

The bending of the flexible cantilever reduces the maximum displacement of the actuator x_a. Thus, the flexibility of the cantilever reduces the pull-in voltage of the actuator as the stiffness of the electrodes increases with increasing distance to the tips. Therefore, the cantilever bending is a compromise between a reduction of the pull-in voltage and a reduction of the maximum displacement. In [19] we show that the maximum displacement range of the actuator x_a is the electrode gap $x_{electrode}$ considering the bending b of the cantilever, which yields:

$$x_a = x_{electrode} - b \tag{1}$$

2.2. System Function

The cooperative system setup is shown in Figure 2a [19]. The system consists of multiple electrostatic bending plate actuators (a_1, a_2, ... , a_j) linked in a chain by connecting springs ($k_1, k_2, \ldots, k_j$) with the identical stiffness k. The first actuator a_1 has the smallest initial electrode gap x_1. With each actuator, the initial gap increases, so the last actuator a_j has the largest initial electrode gap x_j. A main sinusoidal guiding spring k_g [20] is directly connected to the last actuator and guarantees a pure translational system displacement $x_{\text{system},i}$. The connecting spring k_1 and the guiding spring k_g are coupled to the substrate of the chip.

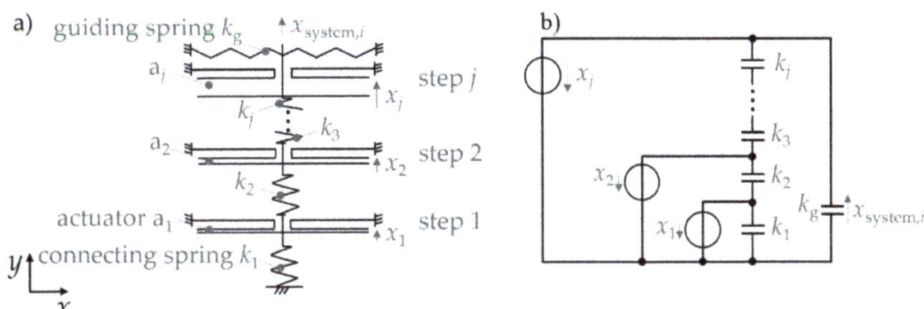

Figure 2. (**a**) Setup of the system, (**b**) analogous electrical network model, (drawings not to scale).

When activating the first actuator a_1, this actuator displaces and its displacement is conducted through the system generating a system displacement $x_{\text{system},1}$. The displacement of the system $x_{\text{system},i}$ when activating actuator a_i is always identical to the reduction of x_j of the last actuator a_j (for $i \leq j$) minus the deformation of the connecting springs. Therefore, considering the bending of the cantilever (Equation (1)), the maximum total system displacement $x_{\text{system},j}$ is the electrode gap distance x_j minus the bending of the cantilever b_j.

2.3. Step-by-Step and Collective Actuation

When the system is actuated step-by-step, an actuator can only be activated if all actuators with smaller electrode gaps are already completely pulled-in. The step-by-step actuation can be described analytically based on an associated analogous electrical network model (Figure 2b). In the network model, the actuators are associated with potential sources and the mechanical springs are modeled as capacitors.

When actuator a_1 is activated, it travels the distance x_1 and the system travels the range $x_{\text{system},1}$. When the actuators a_1 and a_2 are activated, the system travels the range $x_{\text{system},2}$. By successively activating the individual electrostatic actuators, a stepwise system displacement is generated. Therefore, $x_{\text{system},j}$ represents the displacement of the system composed of j actuators when all actuators are displaced and, e.g., $x_{\text{system},3}$ represents the displacement of a system composed of j actuators when the actuators a_1, a_2 and a_3 are displaced (for $3 \leq j$). The system displacement $x_{\text{system},i}$ when activating i actuators in a system consisting of j actuators ($i \leq j$) is described by Equation (2):

$$x_{\text{system},i} = \frac{x_i}{k_g \cdot \left(\frac{j-i}{k_i} + \frac{1}{k_g} \right)} \qquad , \text{ for } i \leq j \tag{2}$$

When applying a voltage to the first actuator a_1, this actuator generates a displacement x_1 that is conducted through the entire system. This initial displacement reduces the gaps between the electrodes of all following actuators a_2 to a_j. For a structure composed of j actuators, Equation (3) describes the displacement x_i of actuator a_i when actuating all actuators with a smaller electrode gap distance ($a_1, a_2, \ldots, a_{i-1}$):

$$x_{a,i} = \sum_{m=2}^{i} x_{\text{system},m-1} \cdot k_g \cdot \left(\frac{j-i}{k_i} + \frac{1}{k_g} \right) , \text{ for } 2 \leq i \leq j \tag{3}$$

Based on Equations (2) and (3), a higher stiffness of the guiding spring k_g reduces the system displacement, whereas a higher stiffness of the connecting springs ($k_1, \dots, k_j$) increases the displacement.

For the step-by-step actuation, each actuator requires its own wiring and bond pad (BP) making the control circuitry complex and enlarging the system's footprint. In Figure 3a 5-step system is shown. Equation (4) describes the relation between the number of steps i and the number of bond pads p:

$$p = 2 \cdot i + 1 \qquad (4)$$

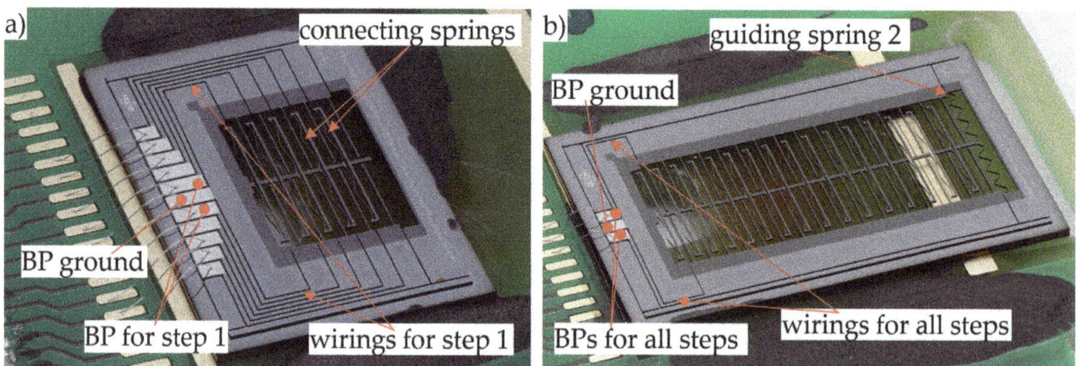

Figure 3. (**a**) 5-step system (system 2) for step-by-step and collective actuation, (**b**) 16-step system (system 6) for collective actuation, (stacked device photos).

During a collective actuation all electrostatic actuators are actuated at the same time so that a single wiring is required on each side, only. The collective actuation is suitable for systems with a large stroke realized by multiple steps, such as the 16-step system shown in Figure 3b. The displacement during collective actuation depends on the applied voltage. Systems with a step-by-step setup can also be actuated collectively, whereas systems with a collective setup cannot be actuated step-by-step.

2.4. System Setup Based on Modelling and Simulation

2.4.1. System Design Based on the Guiding Spring

Sinusoidal guiding springs [20] as shown in Figure 4a are well-suited for the stepwise systems as they feature a very low stiffness in deflection direction lowering the pull-in voltage. However, in wafer plane off-axis direction the high stiffness of these springs ensures a purely translational guiding of the system. Thereby, the selectivity of the sinusoidal spring decreases with an increasing amplitude of the sinusoids [20].

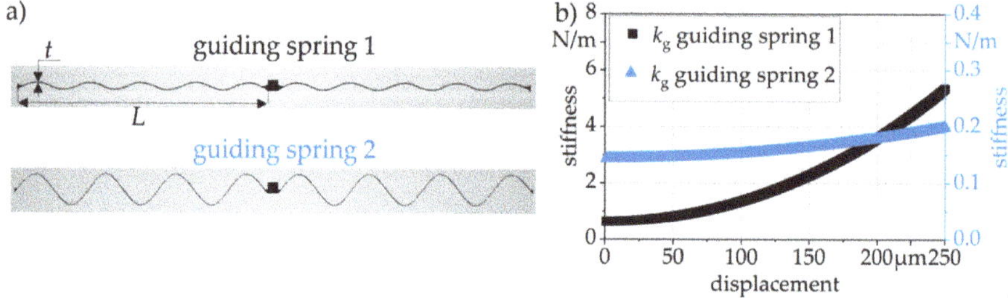

Figure 4. (**a**) Geometry of the guiding springs 1 and 2; (**b**) stiffness of guiding springs 1 and 2 based on solid state COMSOL Multiphysics simulation.

To analyse the influence of the guiding spring on the system function, two different sinusoidal guiding springs are used. Guiding spring 1 features 4.5 sinusoids with an amplitude of 26.8 µm. The length L is 2425 µm, the thickness t amounts to 8 µm, and the width w to 50 µm. Guiding spring 2 has 3.5 sinusoids with an amplitude of 142.25 µm. The length L amounts to 2450 µm, the thickness t is 5 µm, and the width w is 50 µm. The fabricated guiding springs are shown in Figure 15 and the fabricated guiding spring 2 is additionally shown in Figure 3.

Based on COMSOL Multiphysics solid state simulation, the stiffness of guiding spring 1 increases strongly, whereas guiding spring 2 shows a lower and more constant spring stiffness (Figure 4b).

Two system design options can be selected: a design considering a mean stiffness value of the guiding spring and a design considering the actual stiffness value of the guiding spring as a function of displacement.

For both design options, 16-step systems guided by the mentioned guiding spring 1 (Figure 5a) or guiding spring 2 (Figure 5b), respectively, are designed based on Equation (2) to determine the system displacement $x_{\text{system},i}$ and on Equation (3) to determine the electrode gaps x_i. The mean and actual stiffness of the guiding springs are determined based on the COMSOL Multiphysics simulation results presented in Figure 4b.

Table 1. Modeled systems, the simulated mean stiffness of guiding spring 1 is $k_g = 2.9$ N/m and the simulated stiffness of the the connecting springs is $k_i = 4.9$ N/m.

System	Number of Steps	Activation	Design/ Stiffness Value	Guiding Spring	Max. Calculated Displacement	Chip Size [µm × µm]	Presented in Figure
1	5	step-by-step and collective	mean	1	80 µm	6945 × 10,253	6a
2	5	step-by-step and collective	actual	2	94 µm	6945 × 10,253	3a
3	8	collective	mean	2	143 µm	8865 × 9157	6b
4	10	collective	mean	2	171 µm	10,255 × 9157	6b
5	13	collective	mean	2	212 µm	12,392 × 9157	6b
6	16	collective	mean	2	242 µm	14,407 × 9157	3b, 5b, 6b
7	16	collective	actual	1	156 µm	14,407 × 9157	5a

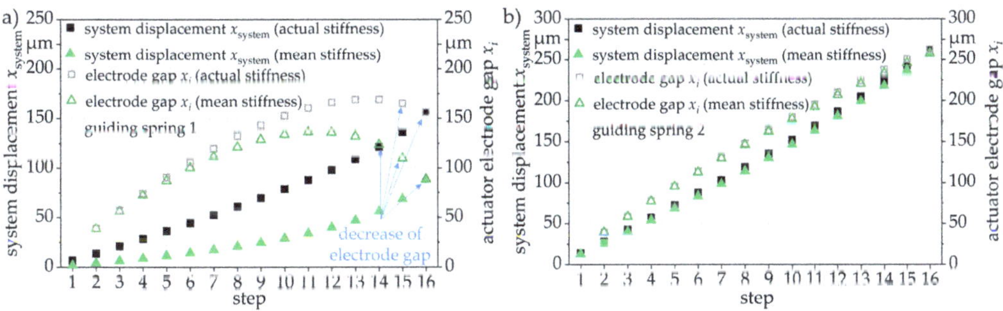

Figure 5. System displacement and electrode gap distance of 16-step systems depending on mean and actual stiffness value design guided by (**a**) guiding spring 1 (actual stiffness value: system 7 presented in Table 1), (**b**) guiding spring 2 (mean stiffness value: system 6 presented in Table 1). The values are obtained analytically (Equations (2) and (3)) by calculating with the spring stiffness values that are obtained by COMSOL simulation. The electrode gap amounts to 20 µm for each actuator.

The step-by-step actuation follows the idea to use the same voltage for the actuation of each single step. The use of the same voltage for each step simplifies the automated electrical control of the actuators with a single voltage source. To ensure that each actuator requires the same voltage, based on Equations (2) and (3), the electrode gap is calculated with the aim to always have the same distance when actuating the actuator. For the calculation an electrode distance of 20 µm is used. Therefore, e.g., for system 1: when no actuator is actuated, actuator a_1 has an electrode gap of 20 µm and actuator a_2 of 38.28 µm. When actuating a_1, actuator a_1 will pull in and based on Equation (3) the actuator a_2 will travel the distance 18.28 µm, so that now the actuator a_2 has an electrode gap of 20 µm. For this setup, we assume that an equal electrode gap results in an equal voltage for complete pull-in and we neglect the cantilever bending.

For steps 1 to 15, the electrode gaps exceed the system displacement as the actuator displacement partly results in a deformation of the connecting springs. Comparing Figure 5a,b the guiding spring design highly influences the system. As the stiffness of guiding spring 2 is almost constant in the required deflection range (Figure 4b), the mean stiffness value and the actual stiffness value systems do not differ much (Figure 5b). Due to the high increase in the stiffness of guiding spring 1, the mean stiffness deviates from the actual spring stiffness so that the actual stiffness value system differs from the mean stiffness value system in Figure 5a. The decrease in the electrode gap as marked in Figure 5a is due to the high stiffness of guiding spring 1 counteracting the system displacement by deforming the connecting springs.

The systems in Figure 5a,b only differ in the guiding spring. Consequently, the guiding spring has a high influence on the resulting system displacement. For both systems, the system displacement is shown based on the actual and on the mean guiding spring stiffness. For the system guided by the low-stiffness spring 2 (Figure 5b), the system displacement does not differ much concerning the actual and the mean stiffness. Thus, the analytical system design is much more complex when calculating with the actual spring stiffness. Therefore, when using a low-stiffness guiding spring, the use of a mean stiffness of the guiding spring can decrease the design effort without a notable influence on the resulting system displacement. For a high-stiffness guiding spring (Figure 5a), it is useful to design the system with the actual spring stiffness to enlarge the resulting system displacement.

2.4.2. System Design for Different Step Numbers

Figure 6 shows electrode gaps of systems featuring a mean stiffness design depending on the number of steps. Based on the idea of equal pull-in voltages as explained in paragraph 2.4.1, the electrode gaps are designed calculating with an electrode gap of 20 µm. Therefore, Equations (2) and (3) are used.

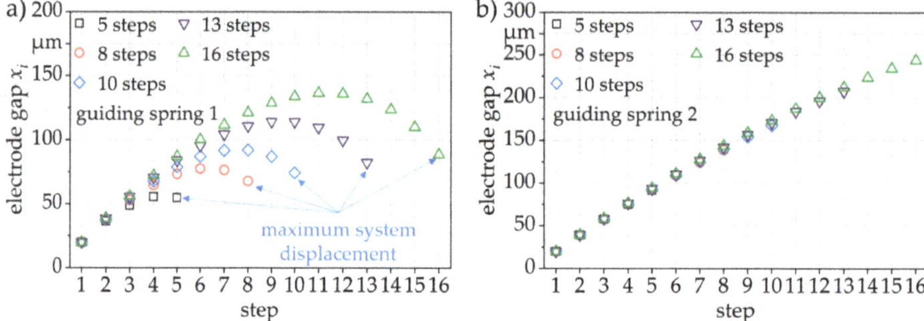

Figure 6. Analytically derived electrode gaps for systems with a mean stiffness value design featuring (**a**) guiding spring 1, (**b**) guiding spring 2.

For the systems in Figure 6a featuring the high stiffness guiding spring 1, the electrode gaps depend on the number of steps in the system. Due to the system setup, the maximum system displacement $x_{\text{system},j}$ is equal to the electrode gap of the last actuator a_j minus the bending b_j of the cantilever. Assuming a negligible small cantilever bending, the calculated total displacements of, the 5-, 8-, 10-, 13-, and 16-step systems amount to 54.7 µm,

67.9 μm, 74.5 μm, 82.5 μm and 88.9 μm, respectively. Consequently, the average step size decreases with an increasing number of steps. The reason is found in the increasing number of connecting springs that deform and therefore reduce the displacement that can be conducted to the following actuators. This deformation of the connecting springs is increased due to the high stiffness of guiding spring 1.

Guiding spring 2 (Figure 6b) features a lower stiffness, and therefore the electrode gaps do not differ much for the systems featuring the low stiffness guiding spring 2. The 5-step system achieves a maximum displacement of 91.9 μm, the 8-, 10-, 13-, and 16-step systems of 139.0 μm, 167.7 μm, 207.5 μm and 244.0 μm, respectively.

2.4.3. Overview of the Modelled Systems

Based on the discussion in paragraph 2.4.1, the systems featuring the low-stiffness guiding spring 2 are modeled with a mean stiffness value (system 3 to 6) and the system featuring the high-stiffness guiding spring 1 with the actual stiffness value (system 7). The only exception are the 5-step systems as the influence of the guiding spring is smaller for a smaller number of steps (paragraph 2.4.2, Figure 6a).

The modeled microsystems are shown in Table 1. All systems feature connecting springs with a stiffness k_i = 4.9 N/m at 20 μm displacement. The fabricated connecting springs are shown in Figure 3. They feature a length of 1965 μm, a thickness of 12 μm, a width of 50 μm and 1.5 sinusoids with an amplitude of 47.5 μm. The systems have 5, 8, 10, 13 and 16 steps. The 5-step systems can be activated step-by-step and collectively. Due to the high number of bond pads, systems with more than 5 steps can only be activated collectively.

3. Fabrication

The microsystems are fabricated on (100)-oriented silicon-on-insulator (SOI) wafers with a 300 μm thick handle layer and a 50 μm device layer.

A 100 nm aluminum layer is deposited on the device layer by evaporation, and the bond pads are patterned by wet chemical etching (Figure 7a). To minimize the risk of a pull-in between the electrostatic actuators and the handle layer during electronic activation, the chips make use of a dicing-free process [21–23] removing the handle layer from the backside of the electrostatic actuators during hydrofluoric (HF) acid vapor etching. This process also separates the single chips from each other, making a mechanical sawing process obsolete. Therefore, the handle layer is deep etched (DRIE, Figure 7b). A 400 nm plasma-deposited SiO_2 layer (Figure 7c) is used as a hard mask for the patterning of the systems. Then, also the device layer is deep etched (Figure 7d). The systems are released by HF-vapor etching (Figure 7e). The electrical isolation of the electrodes is performed by depositing 400 nm silicon nitride (SiN) layer with a low-stress PECVD process (Figure 7f). The low-stress process is required to overcome residual deflections of the very slim electrodes. Single chips are placed on a carrier wafer coated with 50 nm aluminum and flipped to minimize the spacing of the electrodes and the carrying wafer for preventing a parasitic coating of the bond-pads. Finally, the chips are assembled on a printed circuit board (PCB) and wire bonded (Figure 3).

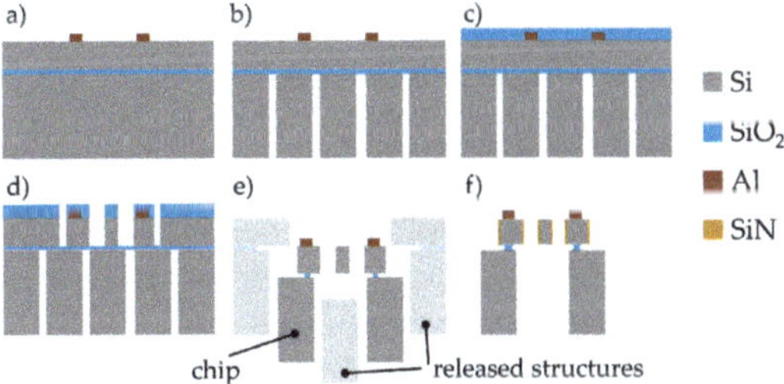

Figure 7. Fabrication process (**a**) etching of bond pads, (**b**) deep-etching of handle layer, (**c**) PECVD of 400 nm SiO_2, (**d**) deep-etching of device layer, (**e**) HF-vapor etching, (**f**) deposition of 400 nm SiN, (drawings not to scale).

4. System Characterization

4.1. Experiment and Characterization Setup

The system performance is evaluated using a highspeed camera system (Keyence VW−600C), the actuators are driven by a voltage source (EA—Electro-Automatic PS-3200-02 C), as presented in Figure 8a. A DC voltage is used for the actuation of the bending plate actuators. The actual system displacement is analyzed with the software *Tracker*. Due to the very high aspect ratio of length vs. displacement and due to the symmetric setup of the system, the videos either investigate the middle, the left or the right side of the system. Experiments are repeated three times under same conditions and averaged.

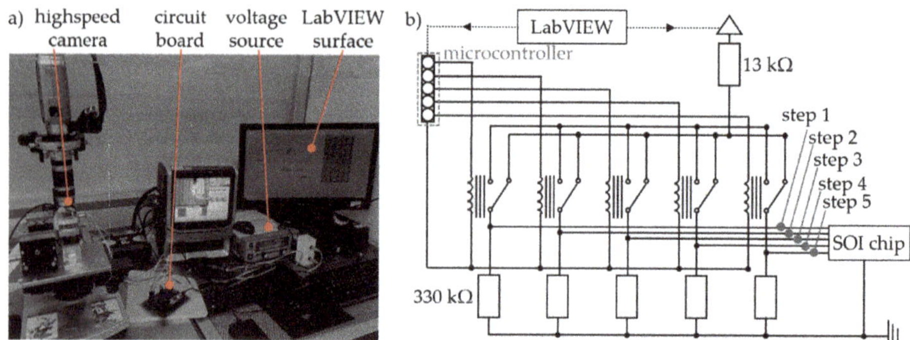

Figure 8. (**a**) Characterization setup in the cleanroom, (**b**) setup of the printed circuit board.

For the collective actuation, the voltage source is driven by a LabVIEW program increasing the voltage as a function of time. A 13 kΩ resistor limits the current flow in case of a potential breakthrough of the SiN isolation layer at the electrodes.

The printed circuit board (PCB) used for the step-by-step actuation is presented in Figure 8b. A LabVIEW program drives the voltage source as well as a microcontroller (Arduino) that activates a set of double pole relays connected to the electrostatic actuators. The combination of double pole relays with the 330 kΩ resistors allows a quick discharge and therefore a quick release of the electrode beams when deactivating the relays. Based on this setup, sticking problems did not appear during the step-by-step actuation. The LabVIEW program allows an activation of all relays as an arbitrary function of time.

4.2. System Characterization Results

4.2.1. Overview of Results

Table 2 provides an overview of the experimental qualitative and quantitative system characterization results that will be explained and discussed in the following chapters.

4.2.2. Design-Based Electrostatic Bending Plate Actuator Behavior

Figure 9 shows the influence of the electrode design on the actuator behavior during activation. In Figure 9a, the electrodes feature the same thickness $c = d = 5$ μm. With increasing voltage, the tips of the electrodes approach due to a bending of the stator electrode. Therefore, the tip of the stator electrode (SE) moves towards the tip of the rotor electrode (RE) (Figure 9b). Then, the electrodes continue their pull-in and start to deform (Figure 9c). The deformation increases with increasing voltage (Figure 9d).

The actuator in Figure 9e features a stator electrode with the thickness $c = 10$ μm and a rotor electrode with the thickness $d = 5$ μm. In this case, with increasing voltage, the rotor electrode bends towards the stator electrode due to its lower stiffness (Figure 9f). Thereby, an electrode deformation is not observed. The higher thickness increases the stiffness of the stator electrode, which seems to prevent the undesired electrode deformation observed with the design of Figure 9a. Due to the electrode deformation, the electrodes could get in contact with other system components placed nearby the actuator, which can have a negative influence on the system function. Therefore, the electrostatic actuators of the stepwise systems are designed as presented in Figure 9e.

Table 2. Overview of the experimental results that will be discussed in the following chapters.

System	Max. Experimental Displacement	Voltage for Max. Displacement	System Properties
1	$x_{system,5} = 71.9 \pm 0.5$ µm	82 V (step-by-step) 71 V (collective)	$x_{system,1} = 5.9$ µm, $x_{system,2} = 22.0$ µm, $x_{system,3} = 38.5$ µm, $x_{system,4} = 55.4$ µm, oscillation during step-by-step actuation
2	$x_{system,5} = 89.4 \pm 0.8$ µm	97 V (step-by-step) 69 V (collective)	$x_{system,1} = 10.6$ µm, $x_{system,2} = 27.9$ µm, $x_{system,3} = 47.3$ µm, $x_{system,4} = 61.4$ µm, oscillation during step-by-step actuation
3	$x_{system,8} = 138.4 \pm 2.3$ µm	74 V	pull-in order: $a_1 + a_2, a_3 + a_4, a_5, a_6, a_7, a_8$
4	$x_{system,10} = 168.3 \pm 3.1$ µm	65 V	pull-in order: $a_1, a_2{-}a_7, a_8, a_9, a_{10}$
5	$x_{system,13} = 202.7 \pm 3.6$ µm	68 V	pull-in order: $a_1, a_2{-}a_4, a_5{-}a_8, a_9, a_{10}{-}a_{12}, a_{13}$
6	$x_{system,16} = 230.7 \pm 0.9$ µm	54 V	pull-in order: $a_1, a_2, a_3, a_4, a_5{-}a_{16}$
7	$x_{system,16} = 138.9 \pm 1.7$ µm	118 V	pull-in order: $a_1, a_2 + a_3, a_4{-}a_6, a_7, a_8, a_9{-}a_{12}, a_{13}, a_{14}, a_{15}, a_{16}$

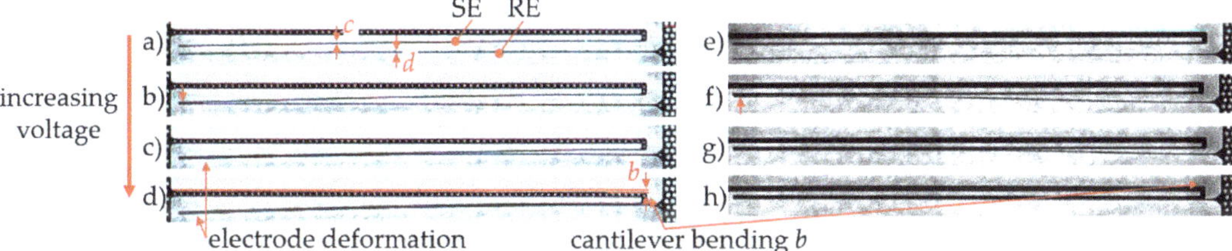

Figure 9. Pull-in depending on electrode thickness, (**a–d**) $c = d = 5$ µm, (**e–h**) $c = 10$ µm, $d = 5$ µm, detailed description in the continuous text.

Figure 10 shows the voltage-dependent experimental displacement of a bending plate actuator and its corresponding cantilever. The voltage is increased in increments of 1 V. Below 30 V, no displacement is observed. At this voltage, the actuator starts displacing, but the cantilever still remains in its position. At 82 V, the cantilever starts to bend down and the pull-in is completed, which can also be seen in Figure 9h. If the cantilever stiffness was higher, the cantilever would not bend and a higher voltage would be required to realize a complete pull-in. During experimental system characterization, the cantilever bending is always the last step leading to a complete pull-in of the bending plate actuators.

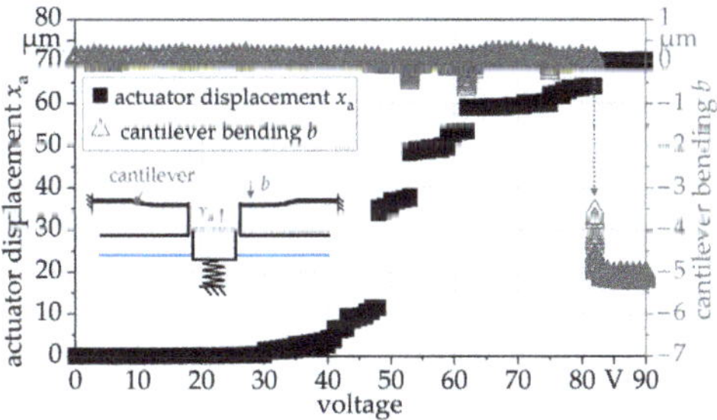

Figure 10. Displacement of the bending plate actuator and bending of the cantilever depending on the applied voltage, for the actuator of step 5 of system 1.

4.2.3. Step-by-Step Actuation of the 5-Step Systems

The step-by-step actuation of system 2 is presented in Figure 11. First, the voltage is increased to 97 V, but all relays on the circuit board remain opened. In Figure 11a, the relays of step 1 are closed which results in a pull-in of step 1. In Figure 11b, the relays of steps 1 to 3 are closed so that the actuators of these steps pull-in. Due to the pull-in of steps 2 and 3, the electrode gaps of steps 4 and 5 decrease. To realize the pull-in of steps 4 and 5, the relays of the corresponding actuators are closed on the PCB (Figure 8b). For the release of the actuators, the relays on the PCB are opened by order again. Thereby, a sticking is prevented due to the double pole relays that are connected to the 330 kΩ resistors, enabling a quick discharge of the electrodes.

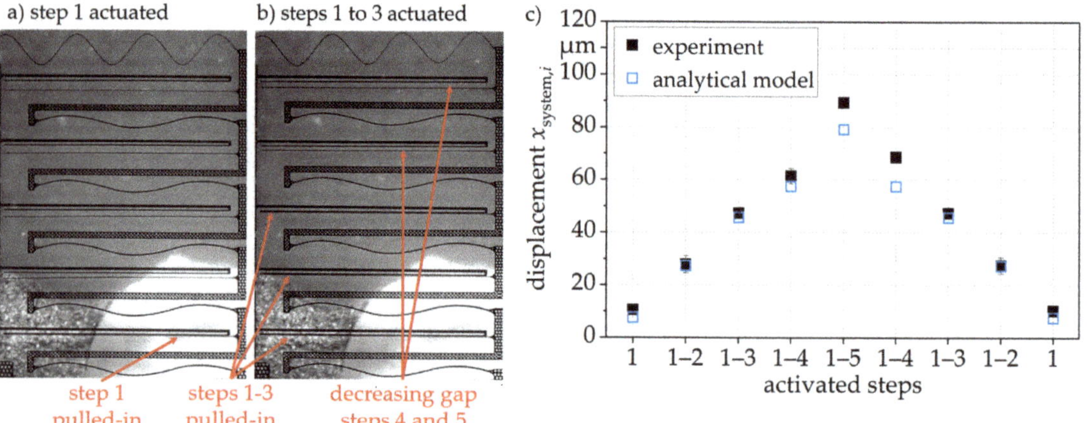

Figure 11. Step-by-step actuation of system 2 (left hand side) at 97 V, relays closed at (**a**) step 1, (**b**) steps 1 to 3, (**c**) experimental and analytical displacement of system 2 at 97 V control voltage, Table 2 presents the exact measured results.

The quantitative evaluation of the stepwise system displacement is shown in Figure 11c presenting the total system displacement $x_{system,5}$, depending on the activated steps. The system shows an approximately linear displacement. A comparison with the calculated displacement shows that the analytical model and the experimental results fit very well. The experimentally derived cantilever bending b differs from step to step. For step 1 it amounts to 11.5 µm, for steps 2 to 5 the cantilever bending amounts to 10.1 µm, 10.1 µm, 15.8 µm and 11.0 µm.

To examine the minimum switching speed of the actuators, the LabView program was used to control the time in which the relays on the PCB (Figure 8b) were opened and closed. Thereby, the actuators were successfully deactivated and activated within 1 ms and videos with 1000 fps were taken. Thereby, no sticking of the electrodes was observed which is attributed to the double pole relays combined to the 330 kΩ resistors placed on the PCB (Figure 8b).

Thus, an oscillation of the systems is observed taking about 0.1 s until the oscillation calms down again. To further examine this oscillation, the stepwise activation presented in Figure 11c was repeated, but this time the displacement of each single step was recorded with 500 fps enabling the tracking of the oscillation. Again, the setup with the PCB was used to open and close the double pole relays, and therefore to charge and discharge the actuators. The double pole relays are connected to the single actuators and controlled by a LabView program enabling a time-dependent control of the relays. The single steps are switched every 0.1 s which corresponds to a frequency of 10 Hz. The time-dependent system displacement is shown in Figure 12. When steps 1 and 2 are pulled in, the oscillation of the system decreases, which is attributed to the stiffness of the system which increases with each pulled-in actuator. Comparing Figure 12a,b, System 1 shows a lower oscillation amplitude, which is ascribed to the higher stiffness of its guiding spring.

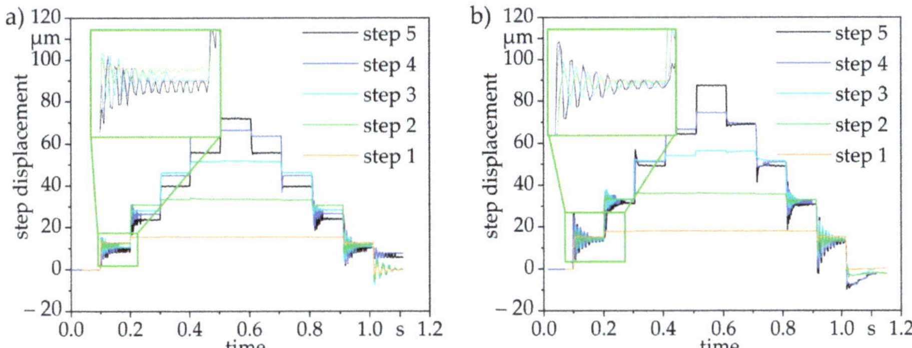

Figure 12. Time-dependent displacement of the single steps at step-by-step activation of (**a**) system 1, (**b**) system 2.

4.2.4. Comparison of Step-by-Step and Collective Actuation

During collective actuation, the system displacement depends on the applied voltage. The pull-in for actuators with smaller initial electrode gaps is achieved first (Figure 13a), and the actuators with larger electrode gaps follow (Figure 13b). The voltage-dependent displacement of the single steps of the 5-step system during collective actuation is shown in Figure 13c. Thereby, all steps start to displace at the same voltage and every step displaces until its pull-in is completed.

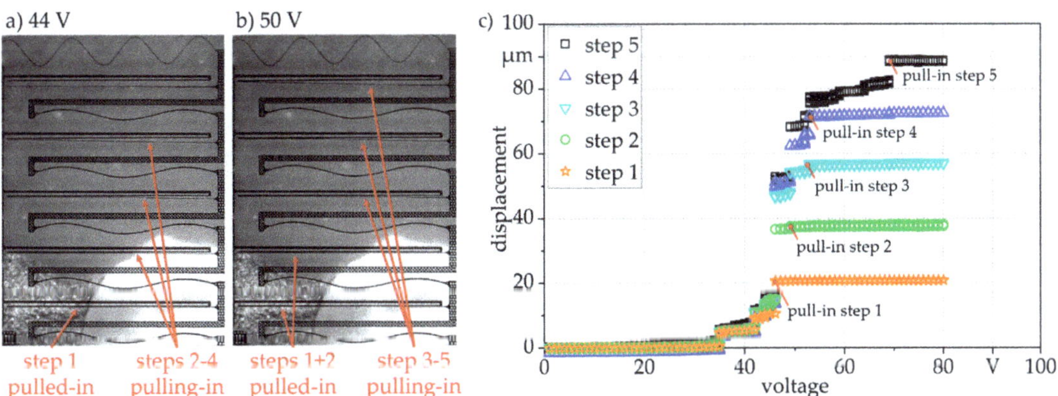

Figure 13. Collective actuation of system 2 at (**a**) 44 V, (**b**) 50 V, (**c**) experimental displacement of the single steps of system 2, the voltage is increased in 1 V-steps every 2 s.

Comparing the step-by-step and the collective actuation, the total system displacement is independent from the mode of activation, contrary to the control voltage. During collective actuation, the electrode tips of the actuators with larger initial electrode gaps already start a movement to each other at low voltage, and therefore reduce the electrode gap of these steps. Consequently, the displacement of the steps with larger initial electrode gaps supports the displacement of steps with smaller initial electrode gaps, resulting in a lower control voltage (Table 2).

4.2.5. Characterization of Microsystems with 5 to 16 Steps

The systems 2 to 6 only differ in the number of steps and their displacement behavior can easily be compared. Figure 14a shows the voltage-dependent displacement during collective actuation. The systems with more steps achieve a higher displacement. Generally, the pull-in voltage is in the same range and varies between 54 V for the 16-step system and 74 V for the 8-step system.

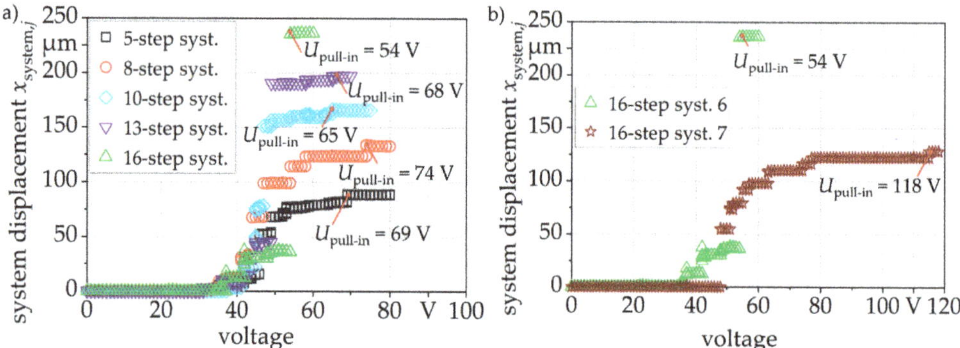

Figure 14. System displacement depending on applied voltage of the 5-, 8-, 10-, 13- and 16-step systems during collective activation, comparison of (**a**) systems 2 to 6, (**b**) system 6 and system 7, the voltage is increased in 1 V-steps every 2 s.

Table 2 and Figures 14b and 15 show the influence of the guiding spring on the total system displacement. The 16-step systems 6 and 7 differ only in the guiding spring. Nonetheless, the system displacement behavior differs considerably: For system 6 (Figure 15a), the steps 1 to 4 are pulled-in at 53 V (Figure 15b) and at 54 V the pull-in of all steps is completed (Figure 15c). For system 7 (Figure 15d), steps 1 to 12 are pulled in at 53 V (Figure 15e), but for the pull-in of all steps a voltage of 118 V is required (Figure 15f).

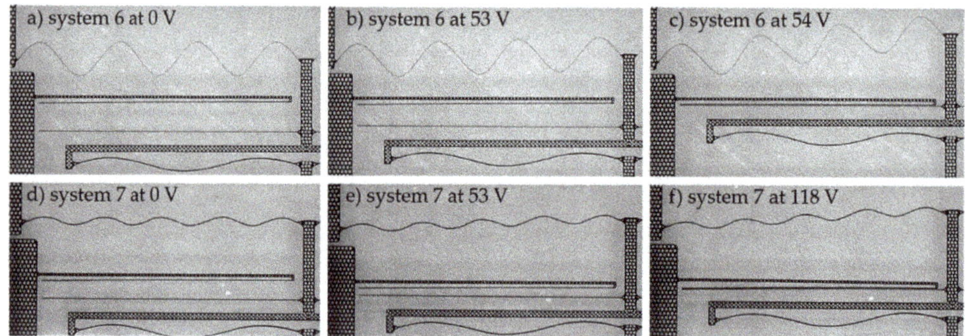

Figure 15. Voltage-dependent total displacement of the 16-step system 6, at (**a**) 0 V, (**b**) 53 V, and (**c**) 54 V and of the 16-step system 7 at (**d**) 0 V, (**e**) 53 V and (**f**) 118 V.

The pull-in of the multi-step systems is further analyzed using video analysis. With increasing voltage, the actuators of the 5-step system pull-in one after each other. For the 8-, 10-, and 13-step systems, the actuators partly pull-in at the same voltage which is presented in Table 2. For system 6, the actuators a_1 to a_4 pull-in in turn with increasing voltage. However, the actuators a_5 to a_{16} pull-in at the same voltage (Figures 14b and 15c). This is attributed to the low stiffness of the guiding spring supporting to conduct an impulse of movement to the following actuators and resulting in a collective pull-in. Furthermore, the high number of interacting electrostatic actuators increases the electrostatic force in the system which is highly facilitating a pull-in.

5. Conclusions

We present the design, fabrication, and experimental characterization of multi-step systems consisting of cooperative electrostatic bending plate actuators and achieving large and stepwise discrete displacements. The systems are modeled based on a derived transfer function with focus on the influence of the guiding spring stiffness on the system function. During experimental characterization, we show the influence of the electrostatic actuator design on the pull-in behavior. By performing step-by-step actuations, we show the fit of our analytic model with the experimental results and we discuss the influence of the guiding spring stiffness on the oscillation of the system. When activating and deactivating the actuators step-by-step, we achieve a minimum switching time of 1 ms. During collective actuation of all electrostatic actuators, the pull-in voltage decreases due to the higher electrostatic force in the microsystem. The 16-step system with low guiding spring stiffness reaches a maximum displacement of 230.7 ± 0.9 μm at 54 V.

Author Contributions: Conceptualization, L.S. and M.H.; methodology, L.S.; software, L.S..; validation, L.S.; formal analysis, L.S.; investigation, L.S.; resources, L.S. and M.H.; data curation, L.S.; writing—original draft preparation, L.S.; writing—review and editing, M.H.; visualization, L.S.; supervision, M.H.; project administration, M.H. and funding acquisition, M.H. All authors have read and agreed to the published version of the manuscript.

Funding: This study was funded by the Deutsche Forschungsgemeinschaft (DFG, German Research Foundation)–Project-ID 287022738–TRR 196, Project C12.

Institutional Review Board Statement: Not applicable.

Informed Consent Statement: Not applicable.

Data Availability Statement: The data can be provided by the author L.S. upon reasonable request.

Acknowledgments: Some segments of the fabrication process were performed at the Center for Interface-Dominated High Performance Materials (ZGH), Ruhr University Bochum.

Conflicts of Interest: The authors declare no conflict of interest. The funders had no role in the design of the study; in the collection, analyses or interpretation of the data; in the writing of the manuscript or in the decision to publish the results.

References

1. Fu, X.; Yang, F.; Liu, C.; Wu, X.; Cui, T.J. Terahertz Beam Steering Technologies: From Phased Arrays to Field-Programmable Metasurfaces. *Adv. Opt. Mater.* **2020**, *8*, 1900628. [CrossRef]
2. Grade, J.D.; Jerman, H.; Kenny, T.W. Design of large deflection electrostatic actuators. *J. Micromech. Syst.* **2003**, *12*, 335–343. [CrossRef]
3. Preetham, B.S.; Lake, M.A.; Hoelzle, D.J. A curved electrode electrostatic actuator designed for large displacement and force in an underwater environment. *J. Micromech. Microeng.* **2017**, *27*, 095009. [CrossRef]
4. Burugupally, S.P.; Perera, W.R. Dynamics of a parallel-plate electrostatic actuator in viscous dielectric media. In *Sensors and Actuators A: Physical*; Elsevier: Amsterdam, The Netherlands, 2019; Volume 295, pp. 366–373.
5. Liu, X.; Samfaß, L.; Kolpatzeck, K.; Häring, L.; Balzer, J.C.; Hoffmann, M.; Czylwik, A. Terahertz Beam Steering Concept Based on a MEMS-Reconfigurable Reflection Grating. *Sensors* **2020**, *20*, 2874. [CrossRef] [PubMed]
6. Hoffmann, M.; Nüsse, D.; Voges, E. Electrostatic parallel plate actuators with large deflections for use in optical moving-fibre switches. *J. Micromech. Microeng.* **2001**, *11*, 323. [CrossRef]
7. Bakri-Kassem, M.; Mansour, R.R. High tuning range parallel plate MEMS variable capacitors with arrays of supporting beams. In Proceedings of the 19th IEEE International Conference on Micro Electro Mechanical Systems, Istanbul, Turkey, 22–26 January 2006; pp. 666–669.
8. Zhang, J.; Zhang, Z.; Lee, Y.C.; Bright, V.M.; Neff, J. Design and invention of multi-level digitally positioned micromirror for open-loop controlled applications. In *Sensors and Actuators A: Physical*; Elsevier: Amsterdam, The Netherlands, 2003; Volume 103, pp. 271–283.
9. Velosa-Moncada, L.A.; Aguilera-Cortés, L.A.; González-Palacios, M.A.; Raskin, J.P.; Herrera-May, A.L. Design of a novel MEMS microgripper with rotary electrostatic comb-drive actuators for biomedical applications. *Sensors* **2018**, *18*, 1664. [CrossRef] [PubMed]

10. Leroy, E.; Hinchet, R.; Shea, H. Multimode hydraulically amplified electrostatic actuators for wearable haptics. *Adv. Mater.* **2020**, *32*, 2002564. [CrossRef] [PubMed]

11. Legtenberg, R.; Groeneveld, A.W.; Elwenspoek, M. Comb-drive actuators for large displacements. *J. Micromech. Microeng.* **1996**, *6*, 320. [CrossRef]

12. Korvink, J.; Oliver, P. *MEMS: A Practical Guide of Design, Analysis, and Applications*; Springer Science & Business Media: New York, NY, USA, 2010.

13. Dean, R.N.; Wilson, C.; Brunsch, J.P.; Hung, J.Y. A synthetic voltage division controller to extend the stable operating range of parallel plate actuators. In Proceedings of the 2011 IEEE International Conference on Control Applications (CCA), Denver, CO, USA, 28–30 September 2011; pp. 1068–1074. [CrossRef]

14. Chiou, J.C.; Lin, Y.C. A novel capacitance control design of tunable capacitor using multiple electrostatic driving electrodes. In Proceedings of the IEEE-NANO 2001, M3.1 Nanoelectronics and Giga-Scale Systems, Maui, HI, USA, 28–30 October 2001; pp. 319–324.

15. Seeger, J.I.; Boser, B.E. Charge control of parallel-plate, electrostatic actuators and tip-in instability. *J. Micromech. Syst.* **2003**, *12*, 656–671. [CrossRef]

16. Lu, M.C.; Fedder, G.K. Position control of parallel-plate microactuators for probe based data storage. *J. Micromech. Syst.* **2004**, *13*, 759–769. [CrossRef]

17. Maithripala, D.H.S.; Berg, J.M.; Dayawansa, W.P. Nonlinear dynamic output feedback stabilization of electrostatically actuated MEMS. In Proceedings of the IEEE Conference on Decision and Control, Maui, HI, USA, 9–12 December 2003; Volume 1, pp. 61–66.

18. Schmitt, L.; Schmitt, P.; Hoffmann, M. 3-Bit Digital-to-Analog Converter with Mechanical Amplifier for Binary Encoded Large Displacements. *Actuators* **2021**, *10*, 182. [CrossRef]

19. Schmitt, L.; Schmitt, P.; Barowski, J.; Hoffmann, M. Stepwise Electrostatic Actuator System for THz Reflect Arrays. In Proceedings of the International Conference and Exhibition on New Actuator Systems and Applications, Online, 17–19 February 2021.

20. Schmitt, P.; Schmitt, L.; Tsivin, N.; Hoffmann, M. Highly Selective Guiding Springs for Large Displacements in Surface MEMS. *J. Micromech. Syst.* **2021**, *30*, 597–611. [CrossRef]

21. Overstolz, T.; Clerc, P.A.; Noell, W.; Zickar, M.; de Rooij, N.F. A clean wafer-scale chip-release process without dicing based on vapor phase etching. In Proceedings of the 17th IEEE International Conference on Micro Electro Mechanical Systems. Maastricht MEMS 2004 Technical Digest, Maastricht, The Netherlands, 25–29 January 2004.

22. Sari, I.; Zeimpekis, I.; Kraft, M. A dicing free SOI process for MEMS devices. *Microelectron. Eng.* **2012**, *95*, 121–129. [CrossRef]

23. Schmitt, L.; Schmitt, P.; Liu, X.; Czylwik, A.; Hoffmann, M. Micromechanical Reflect-Array for THz Radar Beam Steering based on a Mechanical D/A Converter and a Mechanical Amplifier. In Proceedings of the 2020 Third International Workshop on Mobile Terahertz Systems (IWMTS), Essen, Germany, 1–2 July 2020.

Article

Design and Characterization of an Electrostatic Constant-Force Actuator Based on a Non-Linear Spring System

Anna Christina Thewes [1,*], Philip Schmitt [1], Philipp Löhler [1,2] and Martin Hoffmann [1]

[1] Faculty of Electrical Engineering and Information Technology, Chair for Microsystems Technology, Ruhr University Bochum, 44801 Bochum, Germany; Philip.Schmitt@rub.de (P.S.); philipp.loehler@uni-due.de (P.L.); Martin.Hoffmann-mst@rub.de (M.H.)

[2] Chair of Electric Components and Circuits, University of Duisburg-Essen, 47057 Duisburg, Germany

* Correspondence: anna.thewes@rub.de

Abstract: In recent years, tissue engineering with mechanical stimulation has received considerable attention. In order to manipulate tissue samples, there is a need for electromechanical devices, such as constant-force actuators, with integrated deflection measurement. In this paper, we present an electrostatic constant-force actuator allowing the generation of a constant force and a simultaneous displacement measurement intended for tissue characterization. The system combines a comb drive structure and a constant-force spring system. A theoretical overview of both subsystems, as well as actual measurements of a demonstrator system, are provided. Based on the silicon on insulator technology, the fabrication process of a moveable system with an extending measurement tip is shown. Additionally, we compare measurement results with simulations. Our demonstrator reaches a constant-force of $79 \pm 2\,\mu\mathrm{N}$ at an operating voltage of 25 V over a displacement range of approximately $40\,\mu\mathrm{m}$, and the possibility of adjusting the constant-force by changing the voltage is demonstrated.

Keywords: electrostatic; microactuator; microsensor; constant-force; comb drive; stimulation; COMSOL; SOI; MEMS; dicing free

Citation: Thewes, A.C.; Schmitt, P.; Löhler, P.; Hoffmann, M. Design and Characterization of an Electrostatic Constant-Force Actuator Based on a Non-Linear Spring System. *Actuators* **2021**, *10*, 192. https://doi.org/10.3390/act10080192

Academic Editors: Jose Luis Sanchez-Rojas and Manfred Kohl

Received: 30 June 2021
Accepted: 6 August 2021
Published: 11 August 2021

Publisher's Note: MDPI stays neutral with regard to jurisdictional claims in published maps and institutional affiliations.

1. Introduction

This paper introduces an electrostatic constant-force actuator for an application in the analysis of viscoelastic materials [1] or the mechanical stimulation in tissue engineering [2]. This moveable microsystem is fabricated on a silicon-on-insulator (SOI) substrate and consists of an electrostatic comb drive and a non-linear spring system. A non-linear spring system generates a constant restoring force, which enables a constant-force reaction along a specific displacement. Recent research of restoring forces on microstructures, such as Timoshenko curved nanobeams or nonlocal gradient elastic beams, can be found in Reference [3–5]. Additionally, the comb drive generates also a constant voltage controlled actuation force. Due to the linear dependency between the capacitance and the displacement, the actuator position can be measured in real-time.

Yang et al. [6] establish a passive-type constant-force microgripper. In their system, the constant force is realized using a combination of inclined bi-stable beams and straight leaf flexures. The microgripper is designed using two separate parts of actuation and constant force modules. The actuation part is driven by six pairs of electrothermal Z-shaped beams that offer a large displacement at low voltage. The resulting constant force is achieved by a combination of positive- and negative-stiffness mechanisms. Zhang et al. [7] propose a similar design in a macroscopic actuation system. Nevertheless, the actuation force is generated with an external linear actuator. Previous research tends to focus on the constant-force actuation rather than the integrated simultaneous measurement of the actual system displacement, as established in the system presented in this work.

The arrangement of the electrostatic comb drives of the presented constant-force actuator is used to enable the constant actuation force on a large travel range, which is introduced by Legtenberg et al. [8]. Comb drives allow for a simultaneous constant force actuation and capacitance measurement. In this paper, the measurement of the capacitance is used to calculate the actual position of the constant-force actuator.

This work focuses on the design and characterization of a new constant-force actuator. The first section of this paper examines the analytical description of the comb drive and the non-linear spring system. Then, a numerical simulation results in the constant-force chip design.

Noteworthy is the extending and moveable measurement tip, that is necessary to reach a sample with the constant-force system. The free tip is fabricated on the silicon-on-insulator using a dicing free process [9] and a predetermined breaking point [10]. Finally, the experimental setup is described, and the characterization of the different components on the microchip is discussed.

2. Materials and Methods

2.1. Electrostatic Comb Drive Actuator

As Figure 1 demonstrates, the actuator consists of two electrically conductive combs. One comb is mechanically fixed while the other is suspended on an elastic suspension spring. Detailed analytical descriptions of comb drives can be found in Reference [8,11]. The design of the suspension spring is essential in order to eliminate unwarranted angular motion of the rotor [12]. It is designed to have a low stiffness k_y parallel to the comb fingers and a much higher stiffness k_x in the transverse direction. The formula for the capacitance of the comb drive is given by

$$C = \frac{2n\epsilon_0\epsilon_r t}{g}(l_0 + y),\qquad(1)$$

where n is the number of the fingers of the movable comb, ϵ_0 is the dielectric constant, ϵ_r is the relative permittivity, t is the depth of the fingers, g is the gap spacing between the fingers, and $A = t(l_0 + y)$ is the overlapping area with l_0 as the initial comb finger overlap and y as the displacement in y-direction.

Equation (1) indicates that the capacitance is linear to the comb drive displacement in y-direction. The differential measurement method $\Delta C = |C_1 - C_2|$, where the capacitance C_1 increases and C_2 decreases during the system displacement in y-direction, is commonly used to reduce the influence of temperature and environmental changes over time.

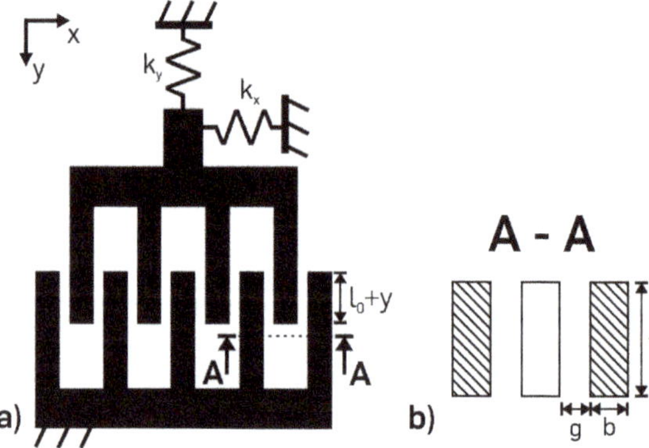

Figure 1. (**a**) Schematic view of a single-sided electrostatic comb drive actuator in y-direction with the finger overlap l_0 and the spring stiffness in x- and y-direction k_x and k_y. (**b**) A to A cross-section of the finger electrodes with finger gap g, width b and depth t.

In case of voltage control, the lateral electrostatic force for the comb drive in the y-direction $F_{el,y}$ can be calculated by

$$F_{el,y} = -\frac{dW_{el}}{dy} = -\frac{n\epsilon_0\epsilon_r t}{g}\cdot U^2,$$ (2)

with U as the applied voltage. From this calculation, it is clear that the electrostatic comb drive force is independent of y and the finger overlap l_0. Because of that, the electrostatic comb drive actuator can generate a constant force that is proportional to U^2.

Ideally, the vertical electrostatic force for comb drive actuators in the x-direction $F_{el,x} = F_{el,x+} - F_{el,x-}$ is almost zero with

$$F_{el,x+} = F_{el,x-} = \frac{n\epsilon_0\epsilon_r t(l_0 + y)}{2(g \pm x)^2}U^2.$$ (3)

However, this ideal theoretical situation does not translate into reality, and large suspension springs are required to suppress the corresponding forces within the mechanical structure. Especially in comb drive actuators, the vertical force causes side instability as a limiting condition. Above a certain voltage, the comb drive operates as a parallel-plate actuator perpendicular to the common movement direction. This unwanted effect requires a careful optimization of the actuator and the mechanical spring for a large displacement [13].

2.2. Non-Linear Spring System

Although the electrostatic actuator generates a constant force, which is independent of the position, the actuator still requires guiding springs, allowing a translational displacement. Serpentine springs [8,14] are commonly used. Considering the ratio of the spring rate in axial and off-axial direction, triangular springs [15] are also well suited. However, by implementing linear springs, the constant force characteristic of the actuator system can be affected, since the restoring force is usually an almost linear function of displacement. For a constant-force system, the force-displacement characteristic of the guiding springs should be constant, as well. A constant force-displacement characteristic can be tailored by combining a mechanical anti-spring featuring a negative spring rate with a linear spring, as indicated in Figure 2.

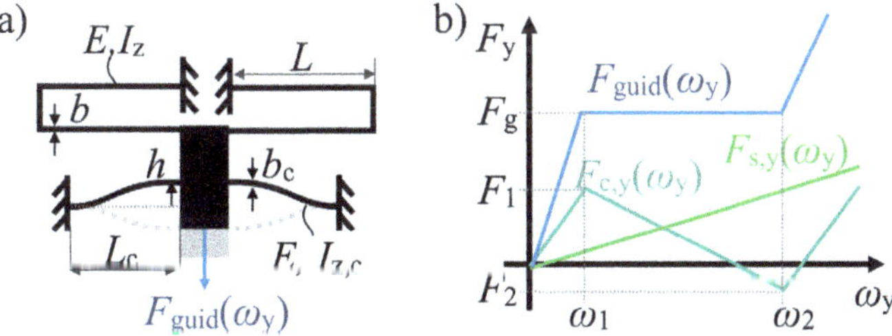

Figure 2. (**a**) Schematic of the combined spring system featuring a linear serpentine spring and a curved cosine-shaped buckling beam. (**b**) The individual and combined force-displacement characteristics of the serpentine and the curved spring.

Since the restoring force of the combined spring system is the superposition of both springs, the negative and positive spring rate compensate each other, resulting in a constant and position-independent force reaction. Considering a serpentine spring as shown in Figure 2 and regarding small displacements in relation to spring size, the restoring force of the spring is known to be:

$$F_{s,y}(\omega_y) = k_{s,y}\omega_y = \frac{12EI_z}{L^3}\omega_y$$ (4)

The expression is based on Euler-Bernoulli beam theory, which also shows high validity for beams realized in monocrystalline silicon. For the design of an anti-spring, a cosine-shaped beam is used, which is described by:

$$\omega(x) = \frac{h}{2}\left(1 - \cos\left(\frac{2\pi x}{L_c}\right)\right). \tag{5}$$

The force-displacement characteristic of the mechanical anti-spring can be calculated by buckling beam analysis. For curved beams, which fulfill the condition of $h/b \geq 6$, Qiu et al. [16] propose an approximation for the calculation of the characteristic positions ω_1, ω_2 and F_1, F_2 of the force-displacement characteristic. Based on these approximations, the force can be written as follows:

$$F_{c,y}(\omega_y) \approx \begin{cases} 578.1\dfrac{EI_{c,z}}{L_c^3}\omega_y & 0 < \omega_y < 0.16h \\[2mm] \dfrac{EI_{c,z}}{L_c^3}\left(\dfrac{185}{2}h - 78.84\omega_y\right) & 0.16h \leq \omega_y < 1.92h \end{cases}. \tag{6}$$

The force-displacement characteristic of the combined spring system finally results in:

$$F_{\text{guid},y}(\omega_y) \approx \begin{cases} \left(\dfrac{12EI_z}{L^3} + 578.1\dfrac{EI_{c,z}}{L_c^3}\right)\omega_y & 0 < \omega_y < 0.16h \\[2mm] \left(1 + \dfrac{12EI_z}{L^3}\right)\left(\dfrac{185}{2}\dfrac{EI_{c,z}}{L_c^3}h - 78.84\dfrac{EI_{c,z}}{8L_c^3}\omega_y\right) & 0.16h \leq \omega_y < 1.92h \end{cases}. \tag{7}$$

The width of both segments in the force-displacement characteristic can only be controlled through the design of the initial deflection of the curved beam. Additionally, a compensation of the positive and negative spring rate is valid for displacements $0.16h \leq \omega_y$. For an ideal compensation, the constant restoring force yields:

$$F_g = 0.16hE\left(\frac{12I_z}{L^3} + 578.1\frac{I_{c,z}}{L_c^3}\right). \tag{8}$$

The force-displacement characteristic is simulated by FEM simulations carried out in COMSOL Multiphysics. A 2D geometry of the non-linear spring system is used for this linear elastic simulation (see Figure 3).

In the solid mechanics interface, the depth of the 2D component is set to 20 μm. The six rounded shoulder fillets are tied to fixed constraints, and a prescribed displacement parameter in y-direction is set. The fine mesh comprises a minimal element size of 0.5 μm, which is more than sufficient for the simulation. In the stationary study, the prescribed displacement parameter varies from 0 to 70 μm by the step size of 1 μm. Taking non-linearities in consideration, the reaction force in y-direction can be simulated. Figure 3 shows the simulation results of a spring system compared to the theoretical results. Both models are performed on the basis of the Young's modulus in [110] direction of monocrystalline silicon and the Poisson's ratio is neglected for simplicity. The chart shows that there are differences between the numerical and analytical constant-force of approximately 10%. This factor may partly be explained by the rounded shoulder fillets of the simulated non-linear spring system to protect the beams from breaking at the clamping. However, the final design of the springs is manually optimized by varying the parameters of the component geometry. In further work, the Nelder-Mead algorithm or the multilevel coordinate search [17] can be used for systematic non-linear optimization.

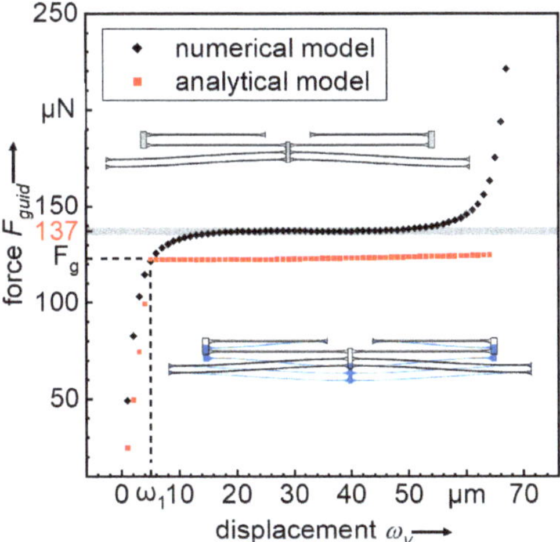

Figure 3. Simulation result of the constant force spring system consisting of one linear serpentine spring and two curved cosine-shaped buckling beams (shown above the graphs). The simulated maximal displaced spring system is pictured under the graphs. The following parameters were used for the analytical model: $E_{[110]} = 169\,\text{GPa}$, $k_{s,y} \approx 3\,\frac{\text{N}}{\text{m}}$, $L_c = 860\,\mu\text{m}$, $b_c = 3.5\,\mu\text{m}$, $h \approx 35\,\mu\text{m}$, $t = 20\,\mu\text{m}$, $I_{z,c} \approx 71.5\,\mu\text{m}^4$. The simulated constant-force is $F_{guid} \approx 137 \pm 2\,\mu\text{N}$ between $\omega_y = 13\,\mu\text{m}$ and $53\,\mu\text{m}$, and the analytical force $F_g \approx 123\,\mu\text{N}$ starts at $\omega_1 = 5\,\mu\text{m}$.

2.3. Design of the Constant-Force Generator

The proposed constant-force microchip combines an electrostatic actuation and a non-linear spring system (Figure 4). For the electrostatic force generation, the actuator consists of $n_a = 52$ moving combs featuring $n = 62$ fingers each. For the differential capacitance measurement method, there are four differential capacitors on each side of the frame. The gap between the finger electrodes, as well as the width of the fingers themselves, is $4\,\mu\text{m}$. A mechanical barrier stops any undue displacements of the system as a precaution against overextension of the springs. A patterned honeycomb structure of the moveable parts reduces the mass load by approximately 80%. Therefore, the weight force of $F_m \approx 0.58\,\mu\text{N}$ is negligible. For this reason, the constant-force F_y results only from the electrostatic force $F_{el,y}$ and the non-linear spring system force F_{guid}:

$$F_y = F_{el,y} - F_{guid}. \tag{9}$$

Therefore, the non-linear spring force needs to be surpassed by the electrostatic actuation ($F_{el,y} > F_{guid}$) to move the measurement tip. The dimensions of a single chip design are $6.5 \times 7.5\,\text{mm}^2$, and the measurement tip has a length of $l_t = 2\,\text{mm}$. This dimension can be adapted individually for every application by slightly changing the mask design.

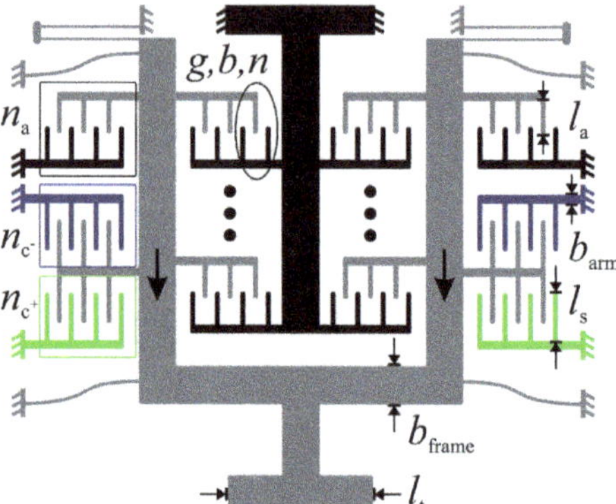

Parameter	Value
g	$4\,\mu m$
b	$4\,\mu m$
n	63
n_a	52
n_{c+}	4
n_{c-}	4
l_a	$110\,\mu m$
l_s	$200\,\mu m$
l_t	$2\,mm$
b_{frame}	$400\,\mu m$
b_{arm}	$30\,\mu m$

Figure 4. Schematic view of the constant-force actuator design that includes fixed combs for the electrostatic actuation and voltage supply (black), moveable combs on a frame with the free standing measurement tip that are suspended by the non-linear spring system on the four corners (grey) and the fixed combs for differential capacitance measurement (blue and green). The parameters $g, b,$ and n are described in Section 2.1.

2.4. Fabrication Process

The demonstrators are fabricated using a SOI-based dicing free process known from, e.g., Reference [9,18]. The main processing steps are shown in Figure 5. A (100)-oriented SOI substrate with $20\,\mu m$ thick device layer and handle layer of $400\,\mu m$ thickness is used. First, a $100\,nm$ thick aluminium layer is deposited on the handle and device layer by evaporation. On the device layer, the aluminium serves as electrode material and is patterned by wet chemical etching (a). In a second step, a SiO_2 hard mask is deposited by PECVD (b). For the release of the chips from the substrate, the aluminium on the handle layer is patterned by wet etching and then etched using deep reactive ion etching (c). In a second deep etching step (e), the mechanical structures on the device layer are etched subsequently after the patterning of the SiO_2 hard mask (d). Next, an HF-vapor process is used to release the mechanical structures by etching the SiO_2 underneath the moveable structures (f). In this processing step, the chips, as well as the backside, are also released from the wafer without dicing (g). Finally, the protection structure for the measurement tip is severed at the predetermined breaking point.

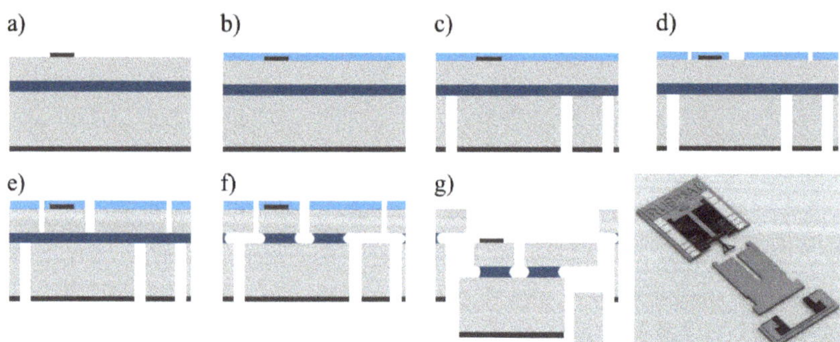

Figure 5. Main fabrication steps of the constant-force chip with free moveable measurement tip: (**a**) aluminium electrodes (**b**) SiO_2 hard mask (**c**) DRIE handle layer (**d**) patterning of the SiO_2 hard mask (**e**) DRIE device layer (**f**) HF-vapor etching (**g**) dicing free.

2.5. Electronical Voltage Control

For the robust and reproducible deflection of the microsystem, an control circuit is developed based on a microcontroller. In addition to the provision of a minimum current amplitude, this circuit has to also provide a reliable and stable voltage supply. To achieve this, the pulse width modulation (PWM) feature of an Arduino Mega 2560 microcontroller is used in combination with a lowpass transistor circuit as shown in Figure 6.

The microcontroller provides an 8-Bit PWM resolution, which results in 256 output voltage values between 0 V and 5 V. This PWM output is then transferred to the base of a NPN transistor that switches a supply voltage U_{dd} between 0 V and 140 V (**1**). In its conducting state, triggered by a base-emitter voltage U_{BE} of 5 V, the supply voltage is attenuated by the voltage divider consisting of the resistors R_3 and R_4. A base-emitter voltage U_{BE} of 0 V leads to the isolation of the transistor and, thus, to a transfer of the entire supply voltage to the next circuit stage. After lowpass filtering, the resulting output voltage U_o (**2**) is given by

$$U_o = U_{dd} - U_{BE} \cdot \frac{R_3}{R_4} \cdot \frac{n_{bit}}{255},\tag{10}$$

where n_{bit} represents the bit value of the microcontroller.

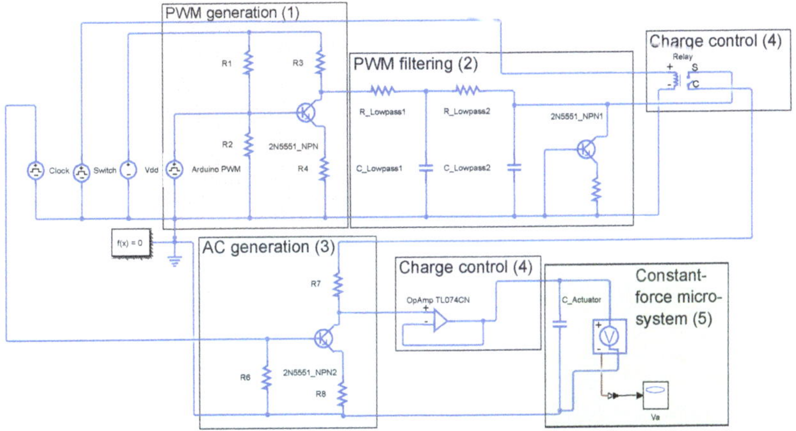

Figure 6. Excitation circuit for the reliable deflection of an electrostatic constant-force microsystem with PWM generation (**1**), PWM filtering (**2**), AC generation (**3**), the charge control components (**4**), and the constant force microsystem (**5**) displayed by its capacitance.

Due to the difference of the spring constants in the x- and y-direction, an overlaid alternating voltage is needed to compensate the lateral displacement of the microactuator elements [19,20]. This voltage is generated by an additional transistor whose base is connected to the clock signal with a variable frequency between 144 Hz and 8 MHz (**3**). The high lowpass resistance leads to a very low collector current into this second transistor, which is in the range of 10 µA to 50 µA. Parasitic effects start to affect the collector-emitter voltage u_{CE} whose usual linear dependency to the base-emitter voltage u_{BE} is disrupted [21]. For a known collector-emitter capacitance C_{CE} in combination with the base-collector resistance r_{BC} and the base-collector capacitance C_{BC}, the relation between those two voltages can be described by

$$\frac{u_{CE}}{u_{BE}} = \frac{1 - (\omega r_{BC} C_{BC})^2}{1 + 4\omega r_{BC} C_{BC} - (\omega r_{BC} C_{BC})^2} \cdot \frac{1}{\sqrt{1 + (\omega C_{CE})}},\tag{11}$$

which corresponds to the product of the transfer functions of a band-stop-filter and a lowpass-filter at ω [22]. The resistances and capacities represent transistor parameters that cannot be modified by external wiring.

To ensure the reproducible deflection of the microsystem, it is necessary to charge the capacitor plates in a shorter time than one period of the systems cutoff frequency. This requires an electrical current amplification of the output signal (**4**) as the limited lowpass output current is unable to deliver a sufficiently large current to the electrodes in time. The current amplification is realized in a buffer amplifier circuit, that provides a maximum current of 1.4 mA, a sufficient amount to charge the capacitor electrodes within nanoseconds. An electromechanical relay provides the required steep voltage rise that leads to the abrupt force generation within the microsystem. In the Appendix A, Table A1 displays the values and types of the electrical components used within the described circuit.

2.6. Experimental Setup for the Force Measurement

The completed microchip is glued onto a printed circuit board (PCB) with a carbon conductive paste for the electrical connection of the microsystem. The aluminium bondpads on the chip in Figure 7a were connected to the contacts on the PCB by bonding with aluminium wire. The handle layer and the moveable parts on the chip are connected to ground for the electric protection of the measured sample. The non-moveable electrostatic combs of the actuation comb drive (C_a) are connected to the input voltage on the PCB. For the differential capacitance measurement, the four sensor pads C_{s1}, C_{s2}, C_{s3}, and C_{s4} were connected separately onto the PCB. The PCB includes pin headers for the electrical connection to voltage control and the capacitive sensing chip on a separate board. This allows for an easy exchange of the constant-force actuators, since all actuators can be controlled with the same electronics.

The capacitance measurement is realized by the *FDC1004Q* capacitance-to-digital converter of *Texas Instruments*. This sensing chip has a measurement resolution of 0.5 fF for an input range of ± 15 pF [23]. Its I^{2}C interface allows the programmable connection with the Arduino Mega 2560 microcontroller. The power supply for the *FDC1004Q* of 3.3 V is also provided by the microcontroller. The FDC1004Q can measure the capacitance between the channel inputs and ground. With its functional mode for differential measurement by

$$\Delta C_1 = C_{s1} - C_{s3} \tag{12}$$

and

$$\Delta C_2 = C_{s2} - C_{s4}, \tag{13}$$

the returned data is almost independent of environmental influences. For the *FDC1004Q*, it is necessary that the capacities C_{s1}, C_{s2} increase and C_{s3}, C_{s4} decrease while the system is moving in y-direction. At all times, these assertions should be fulfilled: $C_{s1} > C_{s3}$ and $C_{s2} > C_{s4}$. For best results, the sensing-chip is located as close as possible to the capacitive sensors. Constant parasitic capacities of the PCB connection can be regarded as a constant capacitive offset.

The experimental setup for force measurement makes use of a high precision balance (*Sartorius WZA 224-N*) and a piezo linear stage (*PI Q-545*). The control of these components is realized in a LabVIEW program that calculates the force from the measured weight with $F = m \cdot g$ and $g = 9.81$ m/s^2. The standard deviation of the balance is given by ± 0.1 mg, which equals approximately 1 µN, and the minimum time to stabilize the measurement is 0.6 s [24]. The time between two measurements is set to 1.5 s. The piezo linear stage has a maximal displacement of 13 mm, a minimal increment of 6 nm and a resolution of 1 nm with a repeatability of 100 nm [25]. As described in Figure 7b, the microchip is fixed to the linear stage that can displace the system in y-direction, positioned vertically to the balance measurement surface. The correct alignment can be controlled optically by a microscope.

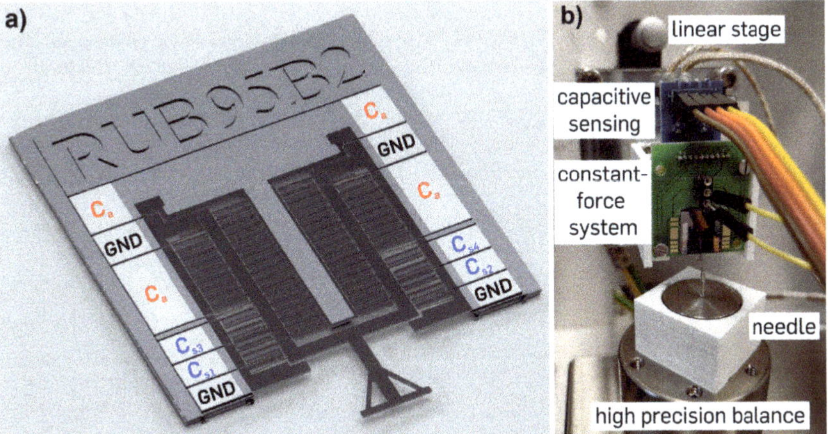

Figure 7. (**a**) Final processed constant-force microchip with electronic connections. (**b**) Experimental setup for the force measurement consists of a piezo linear stage to displace the system in y-direction and a high precision balance.

This setup allows for two different experimental methods: the first method comprises the force measurement of the non-linear spring system without electrical actuation. A needle on the balance is attached to the moveable structure of the chip. When the linear stage moves the PCB upwards step by step, the mass of the needle on the balance is getting lower and the force of the spring system can be measured. The second method is used for the constant-force measurement with electrical actuation. In this case, the needle is directly placed under the measurement tip. Then, the power supply for the comb drive actuation is activated and the resulting constant-force can be measured at the balance. When the stage ascends, the measurement tip exerts pressure on the needle and the resulting force can be measured until the contact to the needle is lost. Simultaneously to the actuation, the deflection can be measured using the capacitance and the sensing-chip.

3. Experiments and Characterization

The constant-force characteristic of the non-linear spring system is shown in Figure 8a (black dots). This measurement graph depicts the average of three force measurements and the standard deviation. When reaching a displacement of 55 μm, a mechanical barrier stops the deflection to protect the non-linear springs. After an initial movement of 8 μm, a constant-force plateau of $F_g = 86 \pm 2$ μN is reached, up to a displacement of 50 μm. For comparison, the analytical and numerical values are depicted in the same graph. The calculated constant-force (red squares) begins at position $\omega_y - 5$ μm, with approximately 102 μN, and the numerical constant-force range (blue diamonds) can be found between $\omega_y = 11$ μm and 56 μm with a constant-force of $F_g = 113 \pm 2$ μN. This results in a measured deviation of approximately 25 % compared to the numerical model and approximately 15% compared to the analytical model. The differences are related to an additional undercut that is generated during the BOSCH-process [26].

The capacitance displacement dependency in comparison to the theoretical calculation is shown in Figure 8b. The graph verifies the linear characteristics of the differential capacitance measurement (Equations (12) and (13)). The right and left capacitances are almost identical, i.e., there is no unwanted twisting of the measurement tip. The maximal displacement of 55 μm implies a measured capacitance of about 1 pF. In theory (Equation (1)) a slope of about 0.022 pF/μm should occur; however, the measured slope is only about 0.017 pF/μm, a difference of approximately 23 %. This can be explained by parameter variations in the fabrication process and additional contact capacities at the PCB board connectors.

Figure 8c shows the results of three measurements of the constant-force system with electrostatic actuation. For the operating voltage U_o, the arduino program was set to $n_{\text{bit}} = 0$, which leads to a voltage drop of $0.6\,\text{V}$ due to the high low pass resistors of $2\,\text{M}\Omega$. The constant-force range can be found between $\omega_y = 5\,\mu\text{m}$ and $45\,\mu\text{m}$ with a force of $F_{\text{const}} = 79 \pm 2\,\mu\text{N}$. The corresponding simulated constant-force range of $40\,\mu\text{m}$ (Figure 3) is confirmed in Figure 8c. Finally, Figure 8d illustrates the force measurement at different operating voltages, and Table 1 includes the measurement results.

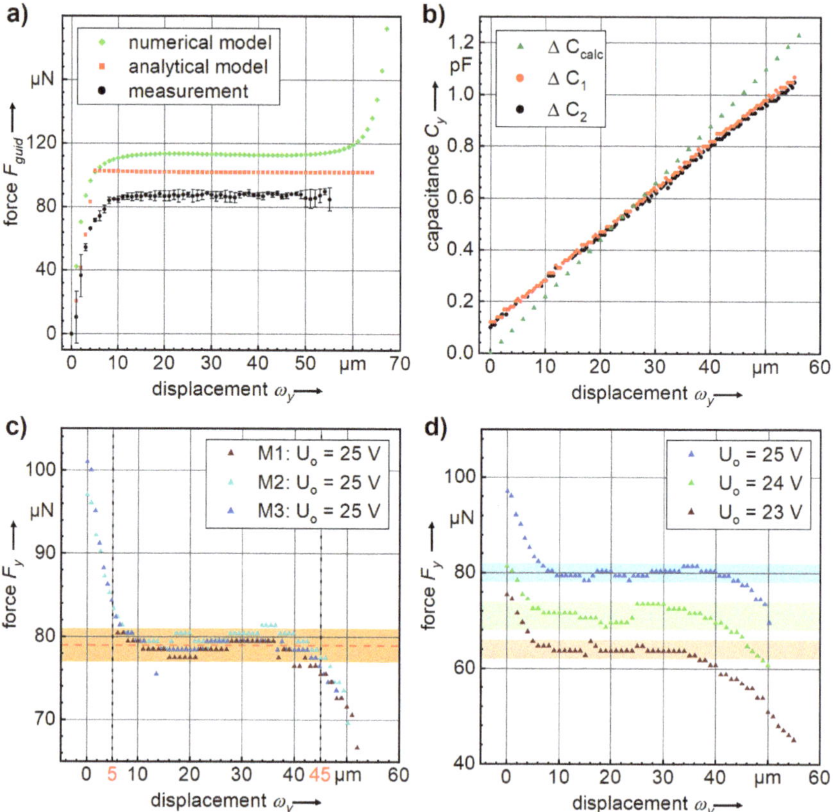

Figure 8. (**a**) Force-displacement characteristic of the non-linear spring system without electrostatic actuation. (**b**) Capacitance-displacement curve measured by the *FDC1004Q* chip and calculated by Formula (1) with $l_0 = 100\,\mu\text{m}$, $n_{c^+} = n_{c^-} = 62$, $g = 4\,\mu\text{m}$, and $t = 20\,\mu\text{m}$. (**c**) Three constant-force measurements with an electrostatic operating voltage $U_o = 25\,\text{V}$ with optimized non-linear spring system parameters $k_{s,y} \approx 3\,\frac{\text{N}}{\text{m}}$ and $b_c = 3.5\,\mu\text{m}$ comparable to Figure 3. (**d**) Constant-force measurement with different operating voltages. For the analytical model, the following parameters were used: $E_{[110]} = 169\,\text{GPa}$, $k_{s,y} \approx 2.47\,\frac{\text{N}}{\text{m}}$, $L_c = 860\,\mu\text{m}$, $b_c = 3.3\,\mu\text{m}$, $h \approx 35\,\mu\text{m}$, $t = 20\,\mu\text{m}$, $I_{z,c} \approx 59.9\,\mu\text{m}^4$.

Table 1. Results of the constant-force measurement at different operating voltages.

Operating Voltage [V]	Constant-Force [μN]	Displacement Range [μm]
23	64 ± 2	35
24	71 ± 3	37
25	79 ± 2	40

As visible in the graph, the constant-force range is shrinking only slightly with decreasing voltage. This allows to adjust the constant-force by changing the operating voltage. The minimum biasing voltage is given by the electrostatic force required to overcome the non-linear spring force. In the case of the presented constant-force actuator, the minimum biasing voltage limit is between 15 and 20 V, depending on the non-linear spring system and the parameter variations in the fabrication process. The maximum biasing voltage is currently limited by the specifications of the transistors and the operational amplifier. In this case, the transistors tolerate voltages up to 140 V, while the op-amp has a maximum input voltage of 35 V.

To summarize, these measurements demonstrate the generation of a static constant-force with the presented microsystem. To generate a dynamic force, an alternating voltage needs to be applied to the constant-force actuator. The maximum frequency of this dynamic force is limited by the resonance frequency of the microsystem, which is located around 50 Hz.

4. Conclusions

This paper presents an electrostatic constant-force actuator with a large displacement, as well as a mathematical model for the optimization of the required non-linear spring system. Furthermore, we demonstrate a comb drive structure, which does not only move the extended measurement tip, but is also capable of determining its actual displacement.

The final design of the constant-force microchip results from the optimization with COMSOL simulations. The chips are fabricated on a SOI substrate in common microsystem fabrication technology. Our demonstrator achieves a constant-force between approximately 64 and 79 µN at a operating voltage from 23 to 25 V and a displacement range of 35 to 40 µm. Additionally, we demonstrate the possibility of adjusting the constant-force by changing the operating voltage.

In future research, the fabrication processes should be optimized to compensate the deviation from theory caused by undercut. Some of our tested constant-force microchip designs exhibit problematic side instability, which could be improved by using more than two curved cosine-shaped buckling beams at different parts of the frame. Furthermore, we will focus on experiments with cells or viscoelastic materials and also optimize our system performance.

Author Contributions: Conceptualization, A.C.T., P.S., P.L. and M.H.; methodology, A.C.T., P.S. and M.H.; software, A.C.T., P.S. and P.L.; validation, A.C.T., P.S., P.L. and M.H.; formal analysis, A.C.T. and P.S.; investigation, A.C.T. and P.L.; resources, M.H.; data curation, A.C.T. and P.S.; writing—original draft preparation, A.C.T., P.S. and P.L.; writing—review and editing, A.C.T., P.S., P.L. and M.H.; visualization, A.C.T., P.S. and P.L.; supervision, M.H.; project administration, M.H.; funding acquisition, M.H. All authors have read and agreed to the published version of the manuscript.

Funding: This research received no external funding.

Institutional Review Board Statement: Not applicable.

Informed Consent Statement: Not applicable.

Data Availability Statement: The data can be provided by the author A.T. upon reasonable request.

Acknowledgments: The authors would like to thank M. Jondral, H. Austenfeld, L. Znajdova, and all other colleagues at the cleanroom who contributed to this article. This microsystem have been fabricated in the cleanroom of the Chair for Microsystems Technology and at the Center for Interface-Dominanted High Performance Materials (ZGH), Ruhr University Bochum.

Conflicts of Interest: The authors declare no conflict of interest.

Abbreviations

The following abbreviations are used in this manuscript:

PWM	Pulse width modulation
SOI	Silicon-on-insulator
PCB	printed circuit board

Appendix A. Electrical Components

Table A1. Values and types of used electrical components.

Component	Value	Type
R_1	47 kΩ	-
R_2	1 kΩ	-
R_3	4.7 kΩ	-
R_4	1 kΩ	-
NPN transistor	-	ON 2N5551 160 V bipolar junction transistor
$R_{Lowpass1}$	1 MΩ	-
$C_{Lowpass1}$	10 nF	-
$R_{Lowpass2}$	1 MΩ	-
$C_{Lowpass2}$	10 nF	-
Relay	-	KY-019 5V magnetic relay
R_6	100 kΩ	-
R_7	470 Ω	-
R_8	1 MΩ	-

References

1. Bueche, F. The Viscoelastic Properties of Plastics. *J. Chem. Phys.* **1954**, *22*, 603–609. [CrossRef]
2. Fahy, N.; Alini, M.; Stoddart, M.J. Mechanical stimulation of mesenchymal stem cells: Implications for cartilage tissue engineering. *J. Orthop. Res.* **2017**. [CrossRef] [PubMed]
3. Zhang, P.; Qing, H.; Gao, C.F. Exact solutions for bending of Timoshenko curved nanobeams made of functionally graded materials based on stress-driven nonlocal integral model. *Compos. Struct.* **2020**, *245*, 112362. [CrossRef]
4. Pinnola, F.; Faghidian, S.A.; Barretta, R.; de Sciarra, F.M. Variationally consistent dynamics of nonlocal gradient elastic beams. *Int. J. Eng. Sci.* **2020**, *149*, 103220. [CrossRef]
5. Apuzzo, A.; Barretta, R.; Faghidian, S.; Luciano, R.; de Sciarra, F.M. Nonlocal strain gradient exact solutions for functionally graded inflected nano-beams. *Compos. Part B Eng.* **2019**, *164*, 667–674. [CrossRef]
6. Yang, S.; Xu, Q. Design and simulation of a passive-type constant-force MEMS microgripper. In Proceedings of the 2017 IEEE International Conference on Robotics and Biomimetics (ROBIO), Macao, China, 5–8 December 2017; pp. 1100–1105
7. Zhang, X.; Wang, G.; Xu, Q. Design, Analysis and Testing of a New Compliant Compound Constant-Force Mechanism. *Actuators* **2018**, *7*, 65. [CrossRef]
8. Legtenberg, R.; Groeneveld, A.W.; Elwenspoek, M. Comb-drive actuators for large displacements. *J. Micromech. Microeng.* **1996**, *6*, 320–329. [CrossRef]
9. Sari, I.; Zeimpekis, I.; Kraft, M. A dicing free SOI process for MEMS devices. *Microelectron. Eng.* **2012**, *95*, 121–129. [CrossRef]
10. Goj, B. Entwicklung Eines Dreiachsigen Taktilen Mikromesssystems in Siliicum-Technologie. Ph.D. Thesis, Technische Universität Ilmenau, Ilmenau, Germany, 2014.
11. Lehner, G. Die Grundlagen der Elektrostatik. In *Elektromagnetische Feldtheorie*; Springer: Berlin/Heidelberg, Germany, 2010; pp. 43–117._2. [CrossRef]
12. Leondes, C.T. *MEMS/NEMS: Handbook Techniques and Applications*; Springer: Berlin/Heidelberg, Germany, 2006.
13. Chen, Y.C.; Chang, I.C.M.; Chen, R.; Hou, M.T.K. On the side instability of comb-fingers in MEMS electrostatic devices. *Sensors Actuators A Phys.* **2008**, *148*, 201–210. [CrossRef]
14. Schomburg, W.K. *Introduction to Microsystem Design*; Springer: Berlin/Heidelberg, Germany, 2015. [CrossRef]
15. Schmitt, P.; Schmitt, L.; Tsivin, N.; Hoffmann, M. Highly Selective Guiding Springs for Large Displacements in Surface MEMS. *J. Microelectromech. Syst.* **2021**, 1–15. [CrossRef]
16. Qiu, J.; Lang, J.; Slocum, A. A Curved-Beam Bistable Mechanism. *J. Microelectromech. Syst.* **2004**, *13*, 137–146. [CrossRef]
17. Nanthakumar, S.; Lahmer, T.; Rabczuk, T. Detection of flaws in piezoelectric structures using extended (FEM). *Int. J. Numer. Methods Eng.* **2013**, *96*, 373–389. [CrossRef]
18. Schmitt, L.; Schmitt, P.; Barowski, J.; Hoffmann, M. Stepwise Electrostatic Actuator System for THz Reflect Arrays. In Proceedings of the ACTUATOR; International Conference and Exhibition on New Actuator Systems and Applications, Online, 17–19 February 2021.
19. Huang, W.; Lu, G. Analysis of lateral instability of in-plane comb drive MEMS actuators based on a two-dimensional model. *Sensors Actuators A Phys.* **2004**, *113*, 78–85. [CrossRef]
20. Wickramasinghe, I.P.M.; Berg, J.M. Lateral stability of a periodically forced electrostatic comb drive. In Proceedings of the 2012 American Control Conference (ACC), Montréal, QC, Canada, 27–29 June 2012. [CrossRef]
21. Tietze, U.; Schenk, C.; Gamm, E. *Halbleiter-Schaltungstechnik*, 16th ed.; Springer: Berlin/Heidelberg, Germany, 2019.
22. Göbel, H. *Einführung in die Halbleiter-Schaltungstechnik*; Springer: Berlin/Heidelberg, Germany, 2019. [CrossRef]

23. Texas Instruments. *FDC1004Q 4-Channel Capacitance-to-Digital Converter for Capacitive Sensing Solutions*; Texas Instruments: Dallas, TX, USA, 2015.
24. Sartorius Mechatronics. *Die WZA-N Serie—OEM Wägezellen*; Sartorius Weighing Technology GmbH: Göttingen, Germany, 2011.
25. Physik Instrumente (PI). *Q-Motion Präzisions-Lineartisch*; Physik Instrumente (PI) GmbH & Co. KG: Karlsruhe, Germany, 2020
26. Schmitt, P.; Hoffmann, M. Engineering a Compliant Mechanical Amplifier for MEMS Sensor Applications. *J. Microelectromech. Syst.* **2020**, *29*, 214–227. [CrossRef]

Article

On the Static Pull-In of Tilting Actuation in Electromagnetically Levitating Hybrid Micro-Actuator: Theory and Experiment

Kirill Poletkin [1,2]

[1] Institute of Microstructure Technology—Karlsruhe Institute of Technology, Hermann-von-Helmholtz-Platz 1, 76344 Eggenstein-Leopoldshafen, Germany; kirill.poletkin@kit.edu or k.poletkin@innopolis.ru

[2] Institute of Robotics and Computer Vision, Innopolis University, 1 Universitetskaya Street, 420500 Innopolis City, Russia

Abstract: This work presents the results of the experimental and theoretical study of the static pull-in of tilting actuation executed by a hybrid levitation micro-actuator (HLMA) based on the combination of inductive levitation and electrostatic actuation. A semi-analytical model to study such a pull-in phenomenon is developed, for the first time, as a result of using the qualitative technique based on the Lagrangian approach to analyze inductive contactless suspensions and a recent progress in the calculation of mutual inductance and force between two circular filaments. The obtained non-linear model, accounting for two degrees of freedom of the actuator, allows us to predict accurately the static pull-in displacement and voltage. The results of modeling were verified experimentally and agree well with measurements.

Keywords: micro-actuators; micro-systems; micro-manipulators; levitation; mutual inductance; electrostatic pull-in; eddy current

Citation: Poletkin, K. On the Static Pull-In of Tilting Actuation in Electromagnetically Levitating Hybrid Micro-Actuator: Theory and Experiment. *Actuators* 2021, 10, 256. https://doi.org/10.3390/act10100256

Academic Editor: Jose Luis Sanchez-Rojas

Received: 29 July 2021
Accepted: 24 September 2021
Published: 29 September 2021

Publisher's Note: MDPI stays neutral with regard to jurisdictional claims in published maps and institutional affiliations.

1. Introduction

Electromagnetic levitation micro-actuators employing remote ponderomotive forces, in order to act on a target environment or simply compensate a gravity force for holding stably a micro-object at the equilibrium without mechanical attachment, have already found wide applications and demonstrated a new generation of micro-sensors and -actuators with increased operational capabilities and overcoming the domination of friction over inertial forces at the micro-scale.

There are number of techniques, which provide the implementation of electromagnetic levitation into a micro-actuator system and can be classified according to the materials used and the sources of the force fields in two major branches: electric levitation micro actuators (ELMA) and magnetic levitation micro-actuators (MLMA). In particular, ELMA were successfully used as linear transporters [1] and in micro-inertial sensors [2,3]. MLMA can be further split into inductive (ILMA), diamagnetic (DLMA), superconducting micro-actuators and hybrid levitation micro-actuators (HLMA) [4], which have found applications in microbearings [5–7], micromirrors [8,9], micro-gyroscopes [10,11], micro-accelerometers [12], bistable switches [13], nanoforce sensors [14], microtransporters [15], microaccelerators [16], micromotors [17–19] and resonators [20].

A wide spectrum of physical principles have been utilized and successfully implemented by using different techniques for microfabrication. However, recently developed 3D microcoil technology [21] together with the integration of a polymer magnetic composite material for flux concentration, allows announcing inductive levitation micro-actuator systems—firstly, as systems with established micro-fabrication process in comparison to the other levitation actuator systems and, secondly, as high-performance systems. As a result of this progress, our group demonstrated the inductive levitation actuator system with the record lowest current consumption [7] around tens of mA. This permits us to avoid using standard bulky high-frequency current amplifiers for exciting the ILMA and to replace them with the integrated control circuit including the signal generator and amplifier and with a size comparable to the size of the micro-actuator system [22].

HLMAs, in which, for instance, the inductive levitation micro-actuator system can be joined with a source of electrostatic field, dramatically increase the capabilities of levitated micro-systems [4] and demonstrate a wide range of different operation modes such as the linear and angular positioning, bi-stable linear and angular actuation and the adjustment of stiffness components, as it was reported in [17,23] and presented by the author at Transducers 2017 [24]. These capabilities open a new very promising perspective to create smart micro-actuator systems with new functional abilities implemented, for instance, by means of the coherent cooperation of distributed microactuators, multistable actuation, mechanical and electromagnetic couplings.

Although the linear pull-in actuation has been comprehensively studied theoretically as well as experimentally [25], the angular pull-in phenomenon of tilting actuation in such HLMAs has not yet been addressed. In this article, in order to fill this gap, a prototype of hybrid levitation micro-actuator was fabricated to experimentally demonstrate and characterize the static pull-in behavior of tilting actuation. Using the qualitative technique based on the Lagrangian approach [26], a semi-analytical model for mimicking such the static pull-in behavior based on two circuit approximation of induced eddy current with proof mass is developed. Upon tilting the levitated disc, the geometrical transformation of eddy current circuit is analyzed by means of finite element approach proposed in [27]. As a result of this analysis was developed a semi-analytical function for calculation mutual inductance and corresponding electromagnetic force and torque by Kalantarov–Zeitlin's method [28]. Thus, the particularity of the developed model is to employ this new derived function for calculation of mutual inductance and to account for two degrees of freedom, namely, linear displacement in the vertical direction and angular displacement of the levitated disc. Through this study we verified successfully the given assumptions for modelling and developed the robust analytical tool to describe the static pull-in actuation of HLMAs.

2. Fabrication and Measurements

In order to comprehensively study the static pull-in of tilting actuation executed by the HLMA, we fabricated a microprototype, as shown in Figure 1a, by using a similar fabrication process as the one reported in work [17]. Namely, the hybrid actuator consists of two structures fabricated independently, namely, the Pyrex structure and the silicon structure, which were aligned and assembled by flip-chip bonding into one device with the dimensions: 9.4 mm × 7.4 mm × 1.1 mm, as shown in Figure 1a.

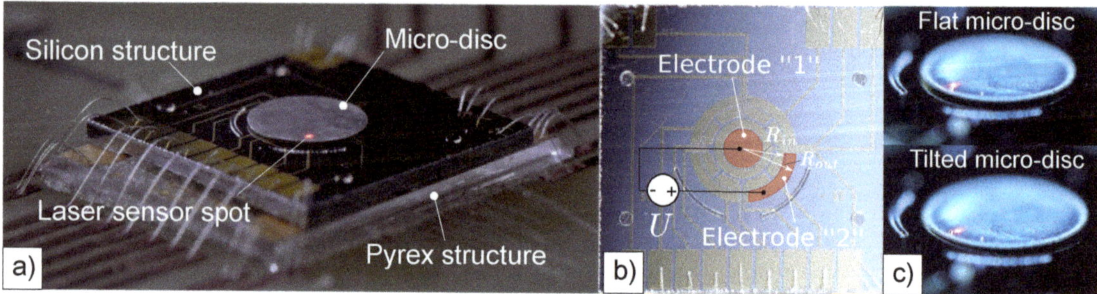

Figure 1. The prototype of micro-actuator-performed tilting pull-in actuation: (**a**) the fabricated prototype of the micro-actuator glued and wire-bonded on a PCB board and levitated stably micro-disc with a diameter of 2.8 mm; (**b**) the set of electrodes fabricated on the top of the silicon structure: the energized electrodes generating the electrostatic forces executing the pull-in actuation are highlighted in red, where *U* is the applied voltage; (**c**) video frames demonstrating actual tilting pull-in actuation: the top figure shows the original flat state of the micro-disc and the bottom figure shows its tilted state after applying the pull-in voltage.

The Pyrex structure includes two coaxial 3D wire-bonded microcoils similar to those reported in our previous work [6], namely, the stabilization and the levitation coils, fabricated on Pyrex substrate using SU-8 2150. The function of this structure is to stably levitate an aluminum disc-shaped proof mass (PM). For this particular device, the height of the coils is 500 μm and the number of windings is 20 and 12 for levitation and stabilization coils, respectively. This coil structure is able to levitate PMs with diameters ranging from 2.7 to 3.3 mm [6].

The silicon structure was fabricated on a SOI wafer with a device layer of 40 μm, the buried oxide of 2 μm, a handle layer of 600 μm and the resistivity of silicon in a range of 1.30 Ω cm, as it was reported in work [29]. Additionally, the device layer has a 500 nm oxide layer for passivation, on top of which electrodes are patterned by UV lithography on evaporated Cr/Au layers (20/150 nm). The fabricated electrode set, which is hidden by the levitated micro-disc in Figure 1a, is shown in Figure 1b. After etching the handle layer up to the buried oxide by DRIE, the silicon structure was aligned and bonded onto the Pyrex structure. Note that the internal structure of the fabricated device is similar to one shown and discussed in our works [17,29]. Finally, the fabricated device was glued and wire-bonded on a PCB board as shown in Figure 1a.

The actuator coils were fed by AC current at the frequency of 10 MHz, with an RMS value ranging from 100 mA to 130 mA. This range of changing of AC current allowed us to levitate a disc with a diameter of 2.8 mm within a height from 35 to 190 μm measured from the surface of the electrodes patterned on the silicon substrate. Pull-in of tilting actuation was performed by applying pull-in voltage to electrodes "1" and "2", as shown in Figure 1b,c. The circular electrode "1" has a diameter of 1 mm corresponding to an area A_1 of around ~0.8 mm^2. Whereas, the geometry of the sectorial electrode "2" is characterized by the following parameters: an inner radius R_{in} of 1.1 mm, an outer radius R_{out} of 1.4 mm and the center angle of $\pi/2$ rad, which corresponds to an area A_2 of around ~0.43 mm^2. The electrode set was able to generate the tilting torque within a range of 0.7×10^{-10} N m. The linear displacement of a disc edge was measured by a laser distance sensor (LK-G32) directly (see Figure 1a,c). The measurements were performed at two levitation heights. Namely, the disc was levitated at 130 and 150 μm. The measured results at these two heights are referred to further in the article as measurement I and II, respectively. The pull-in actuation occurred for the applied voltages of 27 and 33 V and the measured pull-in displacements were 34 and 45 μm, respectively. The results of measurements accompanying the results of modeling are summarized in Table 1 below.

Table 1. Results of measurements and modeling of the static pull-in.

		Measurement I	Measurement II
Measured parameters	Levitation height, h_l	130 μm	150 μm
	Spacing, h	100 μm	120 μm
Results of measurements	**Pull-in displacement**	34 μm	45 μm
	Pull-in voltage, U	27 V	33 V
Parameters for modelling	$\zeta = h_l/d_l$	0.065	0.075
	$\kappa = h/h_l$	0.7692	0.8
Results of medelling	**Pull-in displacement**	38 μm	48 μm
	Pull-in voltage, U	28 V	33 V
Device design	Diameter of levitation coil, d_l	2 mm	
	Area of electrodes, A_1 and A_2	0.8 and 0.43 mm^2	

3. Simulation and Modeling

The mechanism of stable levitation of the disc-shaped proof mass in the framework of two coil design is as follows. The induced eddy currents are distributed along the levitated proof mass in such a way that two circuits with maximum values of eddy current density can be identified. The first circuit corresponds to the eddy current distributed along the edge of disc-shaped PM and the second circuit is defined by the levitation coil. The latter one has a circular path with radius equal to the radius of the levitation coil. Hence, this mechanism can be split into two force interactions. The force interaction occurs between the current in the stabilization coil and induced eddy current corresponding to the first circuit, which contributes mainly to the lateral stability of the levitated PM. Whereas, the force interaction between the current in the levitation coil and induced eddy current related to the second circuit contributes mainly to the vertical and angular stability of the levitated PM [26].

Upon tilting the PM, the eddy current circuit generated by the levitation coil is transformed form a circular shape into an elliptical one. In the section below, this transformation of the shape of eddy current circuit is analyzed by quasi-finite element approach recently developed in [27,30,31]. As a result of the analysis, formulas for calculation of the mutual inductance and corresponding electromagnetic torque and force acting on the PM are derived by employing Kalantarov–Zeitlin's method [28]. Then the derived formulas are applied to the modeling of the pull-in actuation in the HLMA.

3.1. Simulation of Induced Eddy Current within the Tilting Proof Mass

According to the procedure proposed in our previous work [27], the disc is meshed by circular elements, as shown in Figure 2a. For the particular case, the levitated micro-disc is meshed by circular elements, each of them having the same radius of $R_e = 2.4814 \times 10^{-5}$ m. For the disc with a diameter of 2.8 mm, a number of elements is $n = 2496$. 3D scheme of two micro-coils is approximated by a series of circular filaments. The levitation coil is replaced by 20 circular filaments with a diameter of 2.0 mm, while the stabilization coil—by 12 circular filaments with a diameter of 3.8 mm. Thus, the total number of circular filaments, N, is 32. Assigning the origin of the fixed frame $\{X_k\}$ $(k = 1,2,3)$ to the centre of the circular filament corresponding to the first top winding of the levitation coil, the linear position of the circular filaments of levitation coil can be defined as $^{(j)}r_c = [0\ 0\ (j-1) \cdot p]^T$, $(j = 1,\ldots,20)$, where p is the pitch equaling to 25 μm. The same is applicable for stabilization coil, $^{(j)}r_c = [0\ 0\ (j-21) \cdot p]^T$, with the difference that the index j varies from 21 to 32. For both coils, the Bryan angle of each circular filament is $^{(j)}\underline{\phi}_c = [0\ 0\ 0]^T$, $(j = 1,\ldots,32)$. Note that all notations and introduced math variables used in the main text are listed in the nomenclature shown in Appendix A.

The result of meshing becomes a list of elements $\{^{(s)}\underline{C} = [^{(s)}\rho\ ^{(s)}\phi]^T\}$ $(s = 1,\ldots,n)$ containing information about a radius vector and an angular orientation for each element with respect to the coordinate frame $\{x_k\}$ $(k = 1,2,3)$. Now a matrix $\underline{L}$ of self-inductances of circular elements and mutual inductances between them can be formed as follows

$$\underline{L} = L^o\underline{E} + \underline{M}^o, \tag{1}$$

where $\underline{E}$ is the (2496×2496) unit matrix, $\underline{M}^o$ is the (2496×2496)— symmetric hollow matrix whose elements are L^o_{ks} $(k \neq s)$. The self-inductance of the circular element is calculated by the known formula for a circular ring of circular cross-section

$$L^o = \mu_0 R_e\left[\ln 8/\varepsilon_t - 7/4 + \varepsilon_t^2/8(\ln 8/\varepsilon_t + 1/3)\right], \tag{2}$$

where μ_0 is the magnetic permeability of free space, $\varepsilon_t = th/(2R_e)$, th is the thickness of a meshed layer of micro-object (in the particular case, $th = 13$ μm).

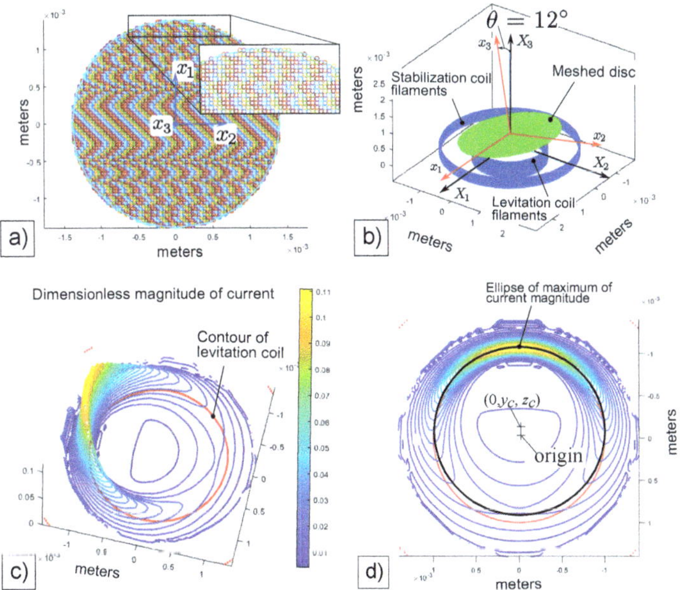

Figure 2. Simulation of eddy current: (**a**) the disc is meshed by circular elements $n = 2496$, $\{x_k\}$ ($k = 1, 2, 3$) is the coordinate frame assigned to the center of disc; (**b**) 3D schematic diagram of the actuator; (**c**) 3D plot of the distribution of dimensionless magnitude of the eddy current along the surface of the disc; (**d**) 2D plot of the distribution of magnitudes of eddy current.

Accounting for the values of diameters of levitation and stabilization coils, 3D geometrical scheme of the actuator for the eddy current simulation can be build as shown in Figure 2b. The position of the coordinate frame $\{x_k\}$ ($k = 1, 2, 3$) with respect to the fixed frame $\{X_k\}$ ($k = 1, 2, 3$) is defined by the radius vector $\boldsymbol{r}_{cm} = [0\ 0\ h_l]^T$, where the levitation height, h_l is to be $100\,\mu\text{m}$. Then, the position of the s-mesh element with respect to the coordinate frame $\{^{(j)}z_k\}$ ($k = 1, 2, 3$) assigned to the j-coil filament can be found as $^{(s,j)}\boldsymbol{r} = \boldsymbol{r}_{cm} + {}^{(s)}\boldsymbol{\rho} - {}^{(j)}\boldsymbol{r}_c$ or in matrix form as

$$^{(s,j)}\underline{r}^z = {}^{(j)}\underline{A}^{zX}\underline{r}_{cm}^X + {}^{(j)}\underline{A}^{zX}{}^{(s)}\underline{\rho}^x - {}^{(j)}\underline{A}^{zX}{}^{(j)}\underline{r}_c^X, \tag{3}$$

where $^{(j)}\underline{A}^{zX} = {}^{(j)}\underline{A}^{zX}\left({}^{(j)}\underline{\phi}_c\right) = {}^{(i)}\underline{e}^z \cdot \underline{e}^X$ and $^{(i)}\underline{A}^{zx} = {}^{(i)}\underline{A}^{zX}\left({}^{(i)}\underline{\phi}_c\right)\underline{A}^{Xx}(\underline{\varphi}) = {}^{(i)}\underline{e}^z \cdot \underline{e}^x$ are the direction cosine matrices, $\underline{\varphi} = [\theta\ 0\ 0]$ is the vector of the angular generalized coordinates.

All matrices of $^{(j)}\underline{\phi}_c$ are zeros. Hence, $^{(j)}\underline{A}^{zX} = \underline{E}$, where $\underline{E}$ is the (3×3) unit matrix and $^{(j)}\underline{A}^{zx} = {}^{(j)}\underline{A}^{zX}\left({}^{(j)}\underline{\phi}_c\right)\underline{A}^{Xx}(\underline{\varphi}) = \underline{E}\,\underline{A}^{Xx}(\underline{\varphi}) = \underline{A}^{Xx}(\underline{\varphi})$. As an illustrative example, the angle θ is chosen to be $12°$, as shown in Figure 2b. Moreover, the coils are represented by the circular filaments and using the radius vector $^{(s,j)}\boldsymbol{r}$, the mutual inductance between the j-coil and s meshed element can be calculated directly using the formula presented in [28]. Thereby, the (2496×32) matrix $\underline{M}_c$ of mutual inductance between coils and finite elements can be formed. The induced eddy current in each circular element is a solution of the following matrix equation

$$\underline{I} = \underline{L}^{-1}\underline{M}_c\underline{I}_c, \tag{4}$$

where $\underline{I}$ is the (2496×1) matrix of eddy currents and $\underline{I}_c = [I_{c1}\,I_{c2}\ldots I_{cN}]^T$ is the given (32×1) matrix of currents in coils.

For calculation, dimensionless currents in the levitation coil and stabilization one are introduced by dividing currents on the amplitude of the current in the levitation coil. Note that the amplitude of the current in both coils are the same. Hence, the input current in the levitation coil filaments is to be one, while in the stabilization coil filaments it is minus one (because of the 180° phase shift). Now, the induced eddy current in dimensionless values can be calculated [27]. The result of calculation is shown in Figure 2c as a 3D plot. The intensity of the color shown by the bar characterizes the value of dimensionless magnitude of the eddy current. Figure 2d shows a 2D plot of the distribution of magnitudes of eddy current along the area of the surface of the PM. As it is seen from Figure 2c,d, the maximum of eddy current magnitudes are distributed along the edge of the PM and around its centre. Moreover, the analysis of Figure 2d depicts that a circular shape of maximum magnitude of eddy current induced by the current of the stabilization coil without tilting the PM [27] is transformed into an elliptical shape with shifting its center, as shown in Figure 2d. The center of the ellipse is defined by the following coordinates $z_c = z_c(\theta)$ and $y_c = y_c(\theta)$, which are also functions of the tilting angle θ.

3.2. Mutual Inductance between Two Filaments of Circular and Elliptic Shapes

Approximating the elliptical shape of distribution of the eddy current magnitude around the centre of the PM by a filament, a function for calculation of the mutual inductance between the elliptical filament with the shifted center and circular filament is derived by using Kalantarov–Zeitlin's method. A scheme with detailed particularities of geometry for calculation is shown in Figure 3.

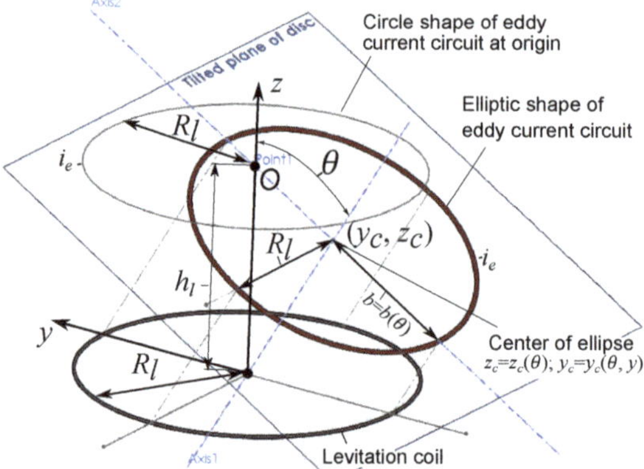

Figure 3. Reduced scheme for modeling electromagnetic interaction between the levitation coil and the tilt-disc: h_l is the levitation height between a plane of coils and equilibrium point; θ is the tilting angle; i_e is the induced eddy current corresponding to the maximum current density within the disc; R_l is the radius of the levitation coil; $z_c = z_c(\theta)$ and $y_c = y_c(\theta, z_c)$ are the coordinates of the center of the ellipse as functions of generalized coordinates; $b = b(\theta)$ is the length of minor axis of the ellipse.

Accounting for Figure 3 and introducing the following dimensionless coordinates,

$$z = \frac{z_c}{R_l}; \; y = -z\frac{\tan(\theta)}{\tan^2(\theta) + 1}, \tag{5}$$

the formula for calculation of the mutual inductance can be written as

$$M = \frac{\mu_0 R_l}{\pi} \int_0^{2\pi} r \cdot F \cdot \Phi(k) d\varphi, \tag{6}$$

where

$$r = r(\theta) = \frac{\cos\theta}{\sqrt{(\tan^2(\theta) + 1)\sin^2(\varphi) + \cos^2(\theta)\cos^2(\varphi)}}, \tag{7}$$

$$F = F(\theta, z) = \frac{R}{\rho^{1.5}} = \frac{r + t_1 \cdot \cos\varphi + t_2 \cdot \sin\varphi}{\rho^{1.5}}, \tag{8}$$

$$\begin{aligned}\rho^2 &= \rho^2(\theta, z) = r^2 + 2r \cdot y\sin(\varphi) + y^2, \\ t_1 &= t_1(\theta, z) = 0.5 \cdot r^2 \tan^2\theta \sin(2(\varphi)) \cdot y, \; t_2 = t_2(\theta, z) = y,\end{aligned} \tag{9}$$

$$\Phi(k) = \frac{1}{k}\left[\left(1 - \frac{k^2}{2}\right)K(k) - E(k)\right], \tag{10}$$

and $K(k)$ and $E(k)$ are the complete elliptic functions of the first and second kind, respectively, and

$$\begin{aligned}k^2 &= k^2(\theta, z) = \frac{4\rho}{(\rho + 1)^2 + z_\lambda^2}, \\ z_\lambda &= \frac{z}{\tan^2(\theta) + 1} + r\tan\theta\sin(\varphi).\end{aligned} \tag{11}$$

3.3. Model of Static Pull-In of Tilting Actuation

It is assumed that the contribution of the force interaction between the current in the stabilization coil and the eddy current circuit flowing along the edge of the PM to the behavior of the PM in the vertical and angular directions measured by the generalized coordinates z_c and θ, respectively, is negligible [27]. Thus, the force interaction between the current in the levitation coil and the eddy current distributed around the PM's center is dominated and determines the actuation of the PM under applying electrostatic force produced by the energized electrodes "1" and "2".

Hence, the Lagrangian of the electromagnetic system under consideration can be written as

$$L = \frac{1}{2}m\dot{z}_c^2 + \frac{1}{2}J\dot{\theta}^2 - mgz_c + M(\theta, z_c)Ii + \frac{1}{2}L_e(\theta, z_c)i^2 - \frac{1}{2}\frac{Q_1^2}{C_1(\theta, z_c)} - \frac{1}{2}\frac{Q_2^2}{C_2(\theta, z_c)}, \tag{12}$$

where m is the mass of PM, J is the moment of inertia of the disc about the axis through center of mass and coinciding with its diameter, g is the gravitational acceleration, $M(\theta, z_c)$ is the mutual inductance defined by Formula (6), I is the given AC current in the levitation coil ($I = \hat{I}e^{j\omega t}$, where $\hat{I}$ is the magnitude, ω is the frequency), i is the induced eddy current, L_e is the self inductance of the eddy current circuit, Q_1 and Q_2 are the charges stored by the planar capacitors C_1 and C_2 built on electrodes "1" and "2", respectively. Accounting for linear and angular displacements of the PM, the capacitances corresponding to circular "1" and sectorial "2" electrode can be described, respectively, by the following equations

$$\begin{aligned}C_1(\theta, z_c) &= -\varphi_{10}\frac{h_l + z_c}{\tan^2(\theta)}\ln\left(1 - \frac{R_1^2\tan^2(\theta)}{(h_l + z_c)^2}\right); \\ C_2(\theta, z_c) &- \frac{\varphi_{20}}{\tan(\theta)}\left[R_{out} - R_{in} + \frac{h_l + z_c}{\tan(\theta)}\ln\left(\frac{1 + R_{in}\tan(\theta)/(h_l + z_c)}{1 + R_{out}\tan(\theta)/(h_l + z_c)}\right)\right],\end{aligned} \tag{13}$$

where $\varphi_{10} = \pi\varepsilon\varepsilon_0$ and $\varphi_{20} = \varepsilon\varepsilon_0\pi/2$, ε is the relative permittivity and ε_0 is the permittivity of free space, R_1 is the radius of the center electrode, R_{out} and R_{in} are the outer and inner radii of the sectorial electrode, respectively. Noting that the capacitors are connected in the series, for charges we can write that $Q_1 - Q_2 = Q$. The dissipation function can be presented as

$$\Psi = \frac{1}{2}R_{eddy}i^2, \tag{14}$$

where R_{eddy} is the electrical resistance of the eddy current circuit. Hence, the state of the system can be defined by two coordinates z_c and θ, the charge Q and the eddy current i.

Adapting the variables and the assumptions introduced above, the Lagrange–Maxwell equations of the system can be written as follows

$$
\begin{cases}
\dfrac{d}{dt}\left(\dfrac{\partial L}{\partial i}\right) + \dfrac{\partial \Psi}{\partial i} = 0; & -\dfrac{\partial L}{\partial Q} = U; \\[3mm]
\dfrac{d}{dt}\left(\dfrac{\partial L}{\partial \dot{z}_c}\right) - \dfrac{\partial L}{\partial z_c} = 0; & \dfrac{d}{dt}\left(\dfrac{\partial L}{\partial \dot{\theta}}\right) - \dfrac{\partial L}{\partial \theta} = 0.
\end{cases}
\tag{15}
$$

Substituting (12) and (14) into (15) and accounting for the fact that the static problem is considered, the acceleration and speed of coordinates z_c and θ are ignored. Thus, we can finally write the following set of equations describing the static behavior of the PM [26]:

$$
\begin{cases}
L_e\dfrac{di}{dt} + R_{eddy}i + M\dfrac{dI}{dt} = 0; & \dfrac{C_2 + C_1}{C_1 C_2}Q = U; \\[3mm]
mg - \dfrac{\partial M}{\partial z_c}Ii - \dfrac{1}{C_1 + C_2}\left(C_2^2\dfrac{\partial C_1}{\partial z_c} + C_1^2\dfrac{\partial C_2}{\partial z_c}\right)U^2 = 0; \\[3mm]
-\dfrac{\partial M}{\partial \theta}Ii - \dfrac{1}{C_1 + C_2}\left(C_2^2\dfrac{\partial C_1}{\partial \theta} + C_1^2\dfrac{\partial C_2}{\partial \theta}\right)U^2 = 0.
\end{cases}
\tag{16}
$$

Since the system operated in the air and the air damping supports the stable levitation of the PM, we can conclude, similar to [26], that $i \approx -MI/L_e$. Substituting the relation for currents from the latter conclusion into (16) and averaging ponderomotive force and torque with respect to the time $(1/2\pi \int_0^{2\pi} \frac{1}{L_e}\frac{\partial M}{\partial p}\Re(i)\Re(I)dt = 1/2\frac{1}{L_e}\frac{\partial M}{\partial p}\hat{I}^2$, where $p = \theta,\ z_c$), we can write

$$
\begin{cases}
mg - \dfrac{1}{2}\dfrac{\partial M}{\partial z_c}\dfrac{M}{L_e}\hat{I}^2 - \dfrac{1}{C_1 + C_2}\left(C_2^2\dfrac{\partial C_1}{\partial z_c} + C_1^2\dfrac{\partial C_2}{\partial z_c}\right)U^2 = 0; \\[3mm]
-\dfrac{1}{2}\dfrac{\partial M}{\partial \theta}\dfrac{M}{L_e}\hat{I}^2 - \dfrac{1}{C_1 + C_2}\left(C_2^2\dfrac{\partial C_1}{\partial \theta} + C_1^2\dfrac{\partial C_2}{\partial \theta}\right)U^2 = 0.
\end{cases}
\tag{17}
$$

4. Analysis of the Derived Model

In order to describe the static pull-in actuation of the fabricated prototype, the semi-analytical model (17) is developed. The developed model (17) is derived by employing the qualitative technique to analyze inductive contactless suspensions, and a new formula for calculation of mutual inductance between the circular filament and its projection on a tilted plane was obtained in Section 3.2. The obtained new formula for calculation of mutual inductance is the result of simulation of induced eddy current within disc-shaped PM received by means of quasi-FEM approach, which showed that the geometry of the second circuit approximating the eddy current induced within the disc by AC current in the levitation coil is transformed due to the disc tilting.

The mechanical behavior of the disc is described by two generalized coordinates. They can be represented by dimensionless variables, namely, $\bar{\theta} = R_T\theta/h$ is the dimensionless angle and $\lambda = z_c/h$ is the dimensionless displacement, where h is the spacing between the surface electrode structure and an equilibrium point O; $R_T = (R_{out} + R_{in})/2$ is the distance between the center of the disc and applied electrostatic force generated by electrode "2". The electromagnetic part of this system is reduced to the interaction between the current in the levitation coil and the induced eddy current with circuit corresponding to its maximum density within the disc. As a result of numerical analysis, the behavior of the shape of induced maximum density of eddy current within the disc was represented via an analytical function, which is dependent on the two generalized coordinates introduced above. This circumstance helps us to exactly define the mutual inductance and then to calculate the magnetic force and torque acting on the disc.

A preliminary result of modeling is shown in Figure 4, which provides an evolution of bifurcation diagrams for equilibrium of the titling PM without linear displacement depending on the applied dimensionless voltage $\sqrt{\beta} = \sqrt{\varepsilon_0 A_1 U^2/(2mgh^2(1+a))}$, where A_1 and A_2 are the areas of electrode 1 and 2, respectively, $a = A_1/A_2$, and depend on dimensionless parameters characterizing geometrical particularities of the prototype design, namely, $\kappa = h/h_l$, which is in a range from 0.5 to 0.9, and $\xi = h_l/(2R_l)$, which changes in a range from 0.03 to 0.09. These ranges of parameters κ and ξ are typical for all known designs of HLMA studied in the literature. As it is seen from Figure 4, increasing a radius of levitation coil (meaning that ξ is decreased) and decreasing the space h (meaning that the dimensionless parameter κ is also decreased) leads to decreasing pull-in voltage and displacement.

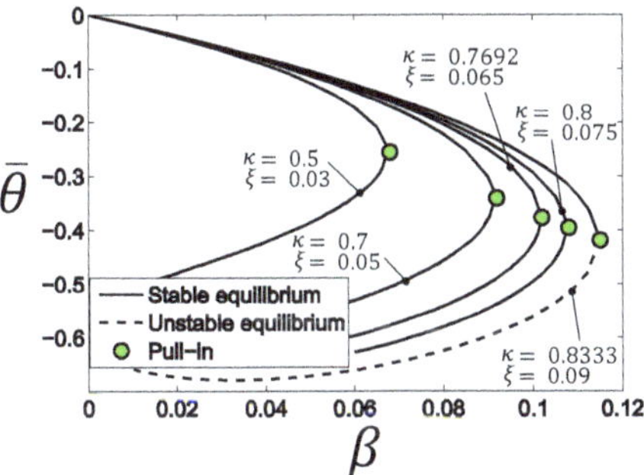

Figure 4. Stable and unstable angular equilibrium of the tilt-disc and its evolution depending on the parameters of device such as $\xi = h_l/2R_l$ and $\kappa = h/h_l$, where h is the space between an electrode plane and equilibrium point of the disc, R_l is the radius of levitation coil.

Modeling the experimental measurement discussed in Section 2 requires taking into account the angular as well as linear displacement of the disc due to essential contribution of linear force acting on the disc along the z-axis. Figure 5 presents the results of modeling and measurements of the static pull-in of tilting actuation executed by the micro-actuator for two different levitation heights, namely, 130 and 150 µm. Data collected during the measurement I and II in absolute values are shown in Figure 5b,d, respectively. Whereas, the results of modeling together with measured data in the normalized values are shown in Figure 5a,c. Analysis of Figure 5 shows that due to the contribution of the linear force, the range of angular displacement is reduced by almost 40% in both measurements.

Furthermore, measured data of pull-in actuation for measurement I and II are shown together with results of modeling in absolute values in Figure 6. Figure 6 provides a conclusive way to estimate the accuracy of the developed model (17) of pull-in actuation in the HLAM by means of direct comparison of results of modeling and measurements. The parameters of the device, particularities of conducted measurements and obtained results are summarized in Table 1. From the analysis of Figures 5 and 6 and Table 1, we can conclude that the results of modeling obtained by means of the developed model are in a good agreement with experimental data.

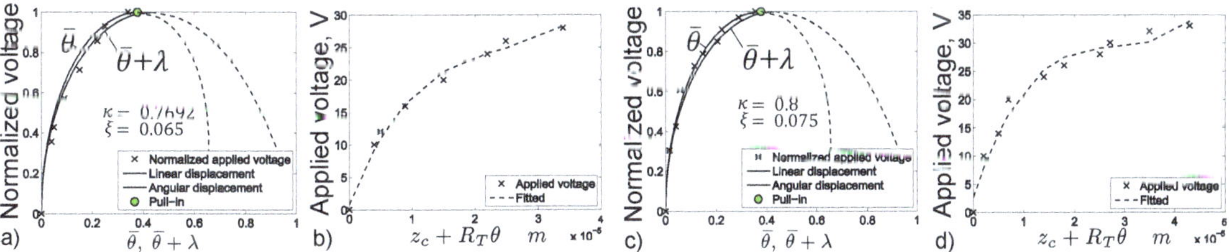

Figure 5. The results of modeling together with measured data in the normalized values: (**a**) measurement I; (**c**) measurement II. Applied voltage vs. the linear displacement of the disc: (**b**) measurement I; (**d**) measurement II (other details are shown in Table 1).

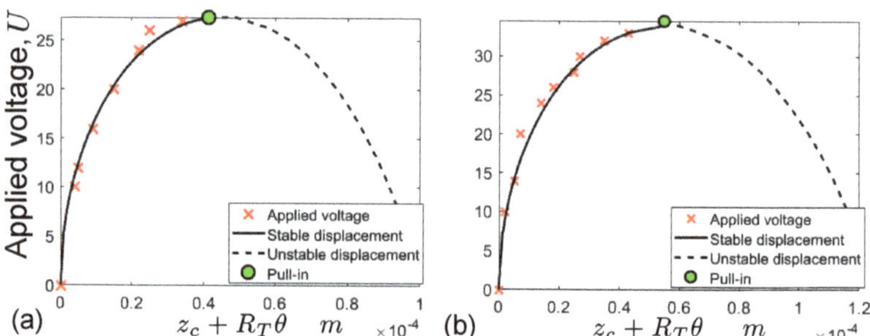

Figure 6. Measured data and results of model=ling in absolute values: (**a**) for measurement I; (**b**) for measurement II (other details are shown in Table 1).

5. Conclusions

In this article, the static pull-in of tilting actuation in HLMA was studied theoretically, as well as experimentally. The semi-analytical model (17) mimicking the static pull-in behavior of tilting actuation in HLMA is developed by employing the qualitative technique to analyze the dynamics and stability of inductive contactless suspensions, and new formula for calculation of mutual inductance between the circular filament and its projection on a tilted plane. As the result of simulation of induced eddy current within disc-shaped PM carried out using a quasi-FEM approach, which showed the distribution of eddy current around the center of a conducting disc, this new formula was developed and presented in the integral form based on Kalantarov–Zeitlin's method. The obtained non-linear model, accounting for two degrees of freedom of the actuator, allows us to predict accurately the static pull-in displacement and voltage for the particular prototype of HLMA. The results of modeling were verified experimentally and agree well with measurements. Through this study we verified successfully the given assumptions for modeling and developed the robust analytical tool to describe the static pull-in actuation and to predict pull-in parameters for HLMAs.

Funding: This research was funded by the German Research Foundation grant number KO 1883/37-1 within the priority program SPP2206 "KOMMMA", project "A 2D array of cooperative hybrid levitation micro-actuators".

Conflicts of Interest: The author declares no conflict of interest.

Abbreviations

The following abbreviations are used in this manuscript:

ILMA   Inductive Levitation Micro-Actuator
HLMA   Hybrid Levitation Micro-Actuator
PM   Proof Mass
MLMA   Magnetic Levitation Micro-Actuator
ELMA   Electric Levitation Micro-Actuator

Appendix A. Nomenclature

A_1 area of the electrode "1" (m²)

A_2 area of the electrode "2" (m²)

a dimensionless parameter A_1/A_2

C_1 capacitance of capacitor build on electrode "1" (F)

C_2 capacitance of capacitor build on electrode "2" (F)

E the complete elliptic function of the second kind

F_l generalized force ($l = 1,2,3$) (N)

$\boldsymbol{g}$ gravity acceleration vector (m/s^2)

h_l height of levitation (m)

h space between the electrode surface and cm of levitated disc (m)

i induced eddy current (A)

I AC current in the levitation coil (A)

$\hat{I}$ magnitude of AC current in the levitation coil (A)

j imaginary unit

K complete elliptic function of the first kind

N number of wire loops

n number of finite elements

L Lagrange function (J)

L_e self-inductance of the eddy current circuit (H)

L_{jj}^c self-inductance of the j-wire loop (H)

L^o self-inductance of the finite circular element (H)

L_{ks}^c mutual inductance between k- and s-finite circular elements (H)

M between two filaments of circular and elliptic shapes (H)

m mass of levitated object (kg)

Q electric charges (C)

R_{eddy} electrical resistance of the eddy current circuit (Ω)

R_e radius of circular element (m)

R_{in} inner radius of sector electrode (m)

R_l radius of levitation coil (m)

R_{out} outer radius of sector electrode (m)

R_T mean distance (m)

th thickness of micro-object (m)

U voltage (V)

y_c coordinate of the centre of the ellipse along the y-axis (m)

z_c coordinate of the centre of the ellipse along the z-axis (m)

Matrices

$\underline{E}$ unit matrix of size $(n \times n)$

$\underline{I}$ matrix of eddy currents of size $(n \times 1)$ (A)

$\underline{I}_c$ matrix of coil currents of size $(N \times 1)$ (A)

$\underline{M}^o$ symmetric hollow matrix of size $(n \times n)$ whose elements are L_{ks}^o $(k \neq s)$ (H)

$\underline{M}_c$ mutual inductance between coils and finite elements of size $(n \times N)$ (H)

Greek symbols

β dimensionless square voltage

θ angular displacement of the levitated disc (rad)

$\bar{\theta}$ dimensionless angle $R_T\theta/h$

κ dimensionless parameter h/h_l

λ dimensionless displacement z_c/h

ζ dimensionless parameter $h_l/(2R_l)$

Ψ dissipation function (W)

ω frequency of AC current (Hz)

References

1. Jin, J.; Yih, T.C.; Higuchi, T.; Jeon, J.U. Direct electrostatic levitation and propulsion of silicon wafer. *IEEE Trans. Ind. Appl.* **1998**, *34*, 975–984. [CrossRef]
2. Murakoshi, T.; Endo, Y.; Fukatsu, K.; Nakamura, S.; Esashi, M. Electrostatically levitated ring-shaped rotational gyro/accelerometer. *Jpn. J. Appl. Phys.* **2003**, *42*, 2468–2472. [CrossRef]
3. Han, F.T.; Liu, Y.F.; Wang, L.; Ma, G.Y. Micromachined electrostatically suspended gyroscope with a spinning ring-shaped rotor. *J. Micromech. Microeng.* **2012**, *22*, 105032. [CrossRef]
4. Poletkin, K.V.; Asadollahbaik, A.; Kampmann, R.; Korvink, J.G. Levitating Micro-Actuators: A Review. *Actuators* **2018**, *7*, 17.
5. Coombs, T.A.; Samad, I.; Ruiz-Alonso, D.; Tadinada, K. Superconducting micro-bearings. *IEEE Trans. Appl. Supercond.* **2005**, *15*, 2312–2315.
6. Lu, Z.; Poletkin, K.; den Hartogh, B.; Wallrabe, U.; Badilita, V. 3D micro-machined inductive contactless suspension: Testing and modeling. *Sens. Actuators A Phys.* **2014**, *220*, 134–143. [CrossRef]
7. Poletkin, K.V.; Lu, Z.; Moazenzadeh, A.; Mariappan, S.G.; Korvink, J.G.; Wallrabe, U.; Badilita, V. Polymer Magnetic Composite Core Boosts Performance of Three-Dimensional Micromachined Inductive Contactless Suspension. *IEEE Magn. Lett.* **2016**, *7*. [CrossRef]
8. Shearwood, C.; Williams, C.B.; Mellor, P.H.; Chang, K.Y.; Woodhead, J. Electro-magnetically levitated micro-discs. In Proceedings of the IEE Colloquium on Microengineering Applications in Optoelectronics, London, UK, 27 February 1996; pp. 6/1–6/3. [CrossRef]
9. Xiao, Q.; Wang, Y.; Dricot, S.; Kraft, M. Design and experiment of an electromagnetic levitation system for a micro mirror. *Microsyst. Technol.* **2019**, *25*, 3119–3128. [CrossRef]
10. Shearwood, C.; Ho, K.Y.; Williams, C.B.; Gong, H. Development of a levitated micromotor for application as a gyroscope. *Sens. Actuators A Phys.* **2000**, *83*, 85–92. [CrossRef]
11. Su, Y.; Xiao, Z.; Ye, Z.; Takahata, K. Micromachined Graphite Rotor Based on Diamagnetic Levitation. *IEEE Electron Device Lett.* **2015**, *36*, 393–395. [CrossRef]
12. Garmire, D.; Choo, H.; Kant, R.; Govindjee, S.; Sequin, C.; Muller, R.; Demmel, J. Diamagnetically levitated MEMS accelerometers. In Proceedings of the IEEE International Solid-State Sensors, Actuators and Microsystems Conference (TRANSDUCERS 2007), Lyon, France, 10–14 June 2007; pp. 1203–1206.
13. Dieppedale, C.; Desloges, B.; Rostaing, H.; Delamare, J.; Cugat, O.; Meunier-Carus, J. Magnetic bistable micro-actuator with integrated permanent magnets. In Proceedings of the 2004 IEEE SENSORS, Vienna, Austria, 24–27 October 2004; Volume 1, pp. 493–496.
14. Abadie, J.; Piat, E.; Oster, S.; Boukallel, M. Modeling and experimentation of a passive low frequency nanoforce sensor based on diamagnetic levitation. *Sens. Actuators A Phys.* **2012**, *173*, 227–237. [CrossRef]
15. Poletkin, K.V.; Lu, Z.; Wallrabe, U.; Korvink, J.G.; Badilita, V. A qualitative technique to study stability and dynamics of micro-machined inductive contactless suspensions. In Proceedings of the 2017 19th International Conference on Solid-State Sensors, Actuators and Microsystems (TRANSDUCERS), Kaohsiung, Taiwan, 18–22 June 2017; pp. 528–531. [CrossRef]
16. Sari, I.; Kraft, M. A MEMS Linear Accelerator for Levitated Micro-objects. *Sens. Actuators A Phys.* **2015**, *222*, 15–23. [CrossRef]
17. Poletkin, K.; Lu, Z.; Wallrabe, U.; Badilita, V. A New Hybrid Micromachined Contactless Suspension With Linear and Angular Positioning and Adjustable Dynamics. *J. Microelectromech. Syst.* **2015**, *24*, 1248–1250. [CrossRef]
18. Xu, Y.; Cui, Q.; Kan, R.; Bleuler, H.; Zhou, J. Realization of a Diamagnetically Levitating Rotor Driven by Electrostatic Field. *IEEE/ASME Trans. Mechatron.* **2017**, *22*, 2387–2391. [CrossRef]
19. Xu, Y.; Zhou, J.; Bleuler, H.; Kan, R. Passive diamagnetic contactless suspension rotor with electrostatic glass motor. *Micro & Nano Lett.* **2019**, *14*, 1056–1059.
20. Chen, X.; Keskekler, A.; Alijani, F.; Steeneken, P.G. Rigid body dynamics of diamagnetically levitating graphite resonators. *Appl. Phys. Lett.* **2020**, *116*, 243505. [CrossRef]
21. Kratt, K.; Badilita, V.; Burger, T.; Korvink, J.; Wallrabe, U. A fully MEMS-compatible process for 3D high aspect ratio micro coils obtained with an automatic wire bonder. *J. Micromech. Microeng.* **2010**, *20*, 015021. [CrossRef]
22. Vlnieska, V.; Voigt, A.; Wadhwa, S.; Korvink, J.; Kohl, M.; Poletkin, K. Development of Control Circuit for Inductive Levitation Micro-Actuators. *Proceedings* **2020**, *64*, 39. [CrossRef]
23. Poletkin, K.; Lu, Z.; Wallrabe, U.; Badilita, V. Hybrid electromagnetic and electrostatic micromachined suspension with adjustable dynamics. *J. Phys. Conf. Ser.* **2015**, *660*, 012005. [CrossRef]
24. Poletkin, K.V. A novel hybrid Contactless Suspension with adjustable spring constant. In Proceedings of the 2017 19th International Conference on Solid-State Sensors, Actuators and Microsystems (TRANSDUCERS), Kaohsiung, Taiwan, 18–22 June 2017; pp. 934–937. [CrossRef]
25. Poletkin, K.V.; Shalati, R.; Korvink, J.G.; Badilita, V. Pull-in actuation in hybrid micro-machined contactless suspension. *J. Phys. Conf. Ser.* **2018**, *1052*, 012035. [CrossRef]
26. Poletkin, K.; Lu, Z.; Wallrabe, U.; Korvink, J.; Badilita, V. Stable dynamics of micro-machined inductive contactless suspensions. *Int. J. Mech. Sci.* **2017**, *131–132*, 753–766. [CrossRef]
27. Poletkin, K.V. Static Pull-In Behavior of Hybrid Levitation Microactuators: Simulation, Modeling, and Experimental Study. *IEEE/ASME Trans. Mechatron.* **2021**, *26*, 753–764. [CrossRef]

28. Poletkin, K.V.; Korvink, J.G. Efficient calculation of the mutual inductance of arbitrarily oriented circular filaments via a generalisation of the Kalantarov-Zeitlin method. *J. Magn. Magn. Mater.* **2019**, *483*, 10–20. [CrossRef]
29. Poletkin, K.V.; Korvink, J.G. Modeling a Pull-In Instability in Micro-Machined Hybrid Contactless Suspension. *Actuators* **2018**, *7*, 11. [CrossRef]
30. Poletkin, K. Simulation of Static Pull-in Instability of Hybrid Levitation Microactuators. In Proceedings of the ACTUATOR; International Conference and Exhibition on New Actuator Systems and Applications 2021, Online, 17–19 February 2021; pp. 1–4.
31. Poletkin, K.V. *Levitation Micro-Systems: Applications to Sensors and Actuators*, 1st ed .; Springer International Publishing: Cham, Switzerland, 2020; p. 145. [CrossRef]

actuators

Article

Bi-Directional Origami-Inspired SMA Folding Microactuator

Lena Seigner [1], Georgino Kaleng Tshikwand [2], Frank Wendler [2] and Manfred Kohl [1,*]

[1]   Institute of Microstructure Technology, Karlsruhe Institute of Technology,
     76344 Eggenstein-Leopoldshafen, Germany; lena.seigner@kit.edu
[2]   Institute of Materials Simulation, Friedrich-Alexander University of Erlangen-Nürnberg (FAU),
     90762 Fürth, Germany; georgino.tshikwand@fau.de (G.K.T.); frank.wendler@fau.de (F.W.)
*   Correspondence: manfred.kohl@kit.edu

Abstract: We present the design, fabrication, and characterization of single and antagonistic SMA microactuators allowing for uni- and bi-directional self-folding of origami-inspired devices, respectively. Test devices consist of two triangular tiles that are interconnected by double-beam-shaped SMA microactuators fabricated from thin SMA foils of 20 µm thickness with memory shapes set to a 180° folding angle. Bi-directional self-folding is achieved by combining two counteracting SMA microactuators. We present a macromodel to describe the engineering stress–strain characteristics of the SMA foil and to perform FEM simulations on the characteristics of self-folding and the corresponding local evolution of phase transformation. Experiments on single-SMA microactuators demonstrate the uni-directional self-folding and tunability of bending angles up to 180°. The finite element simulations qualitatively describe the main features of the observed torque-folding angle characteristics and provide further insights into the angular dependence of the local profiles of the stress and martensite phase fraction. The first antagonistic SMA microactuators reveal bi-directional self-folding in the range of −44° to +40°, which remains well below the predicted limit of ±100°.

Keywords: self-folding origami; microactuation; shape memory alloy foils; micro technology

Citation: Seigner, L.; Tshikwand, G.K.; Wendler, F.; Kohl, M. Bi-Directional Origami-Inspired SMA Folding Microactuator. *Actuators* **2021**, *10*, 181. https://doi.org/10.3390/act10080181

Academic Editor: Wei Min Huang

Received: 22 June 2021
Accepted: 27 July 2021
Published: 3 August 2021

Publisher's Note: MDPI stays neutral with regard to jurisdictional claims in published maps and institutional affiliations.

1. Introduction

The increasing demand on the functionality of microsystems requires new approaches for active control of mechanical, optical, and fluidic components. Most microactuator concepts are limited to one- or two-dimensional workspaces. The concept presented here uses the technique of self-folding, following the ancient art of paper folding, known as origami, by which flat sheets transform into numerous three-dimensional (3D) shapes. The shape change can be achieved by bimorph structures [1], smart composites [2,3], or smart materials such as shape memory alloys (SMAs), responding to an external stimulus such as the environmental temperature or an electrical current. The advantage of origami-inspired designs in engineering is their compact and deployable setup, which has proven its versatility in various macroscopic applications, e.g., airbags [4], wings [5], or tools for minimally invasive surgery [6]. In addition, the actual 3D function could be assigned after the fabrication of the initially flat 2D structures. An overview of the designs and mechanisms of self-folding structures can be found in recent reviews, e.g., [7–9].

SMAs provide high bending moments, low corrosion, and low fatigue. SMA wires have been used to produce a folding motion [10–12] and furthermore can be transformed into SMA coils to generate larger displacements and bending angles [6,13]. The use of SMA sheets or foils can simplify the design and fabrication of folding actuators at the miniature scale [14]. Previous concepts of origami-inspired self-folding SMA sheets have been published in [10,15]. Further relevant research has been performed, e.g., in the field of robogami [16]. Alternative mechanisms for satellite deployment using SMA hinges were presented in [17]. While previous developments focused on the macro scale, we intend to transfer the concept of origami-inspired SMA folding actuation to the micro scale by combining state-of-the-art methods of SMA film engineering and micromachining. The aim of this study is to design, fabricate, and analyze the performance of single and antagonistic SMA microactuators, allowing for uni- and bidirectional folding at miniature scales.

2. Materials and Methods

2.1. Materials Characterization

The outstanding property of SMAs is their thermally induced solid-state phase transformation, from a low-temperature phase (martensite) to a high-temperature phase (austenite) [18]. When an SMA material is deformed in its martensitic state, it undergoes a diffusionless and reversible phase transformation. Restoring its memory shape requires heating above its phase transformation temperature (austenite finish temperature A_f). In this investigation, the base material is a cold-rolled NiTi foil of 20 µm thickness. Aiming at multi-physical simulation of the folding actuation, the material is characterized with regard to its mechanical, thermal, and electrical properties. The differential scanning calorimetry (DSC) measurement shown in Figure 1 provides information on the martensitic transformation temperatures of the as-received material. The phase transformation regime is roughly located between 5 and 62 °C. On cooling, the material undergoes an intermediate R-phase transformation before the transformation to martensite begins at approximately 19 °C. Therefore, at room temperature, the material is in R-phase condition as long as no external load is applied. The reverse transformation from R-phase to austenite is associated with low thermal hysteresis and excellent fatigue life [19,20]. This behavior is reflected in the electrical resistance characteristic in Figure 1b showing the temperature range of folding actuation above room temperature. After the first load cycle, phase transformation is highly reversible.

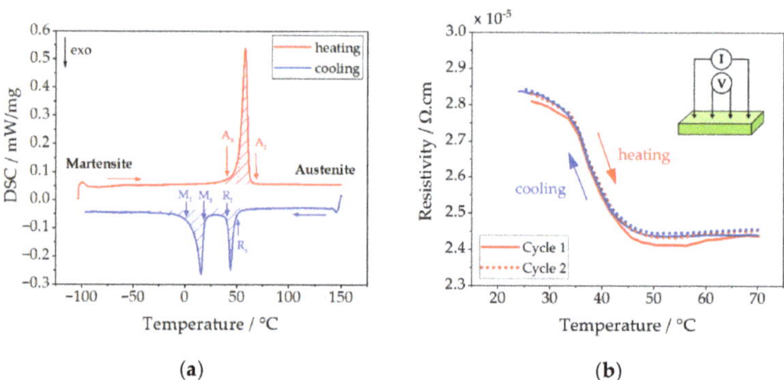

Figure 1. (**a**) Differential scanning calorimetry measurement of a cold-rolled NiTi foil of 20 µm thickness; (**b**) four-point-probe electrical resistivity measurement in the temperature regime of actuation above room temperature.

Tensile tests are carried out in a heating chamber in order to determine the temperature-dependent mechanical parameters including Young's modulus, maximum transformation strain, and critical stress for inter-martensitic transformation. Figure 2a shows experimental and simulated engineering stress–strain characteristics for different temperatures. Typical features are the increase in stiffness and of the critical transformation stress, which is apparent by the shift of the stress plateau toward higher values. The stiffness increase is attributed to the rule of mixture of the involved phase fractions, including the 'soft' R-phase and 'hard' austenite that change with temperature. The critical transformation stress increases linearly following the Clausius Clapeyron relation with coefficient C_{AM}. The experimental characteristic at room temperature exhibits a rather large slope of the stress plateau due to strain hardening caused by sample fabrication. The obtained material parameters are summarized in Table A1 in the Appendix A.

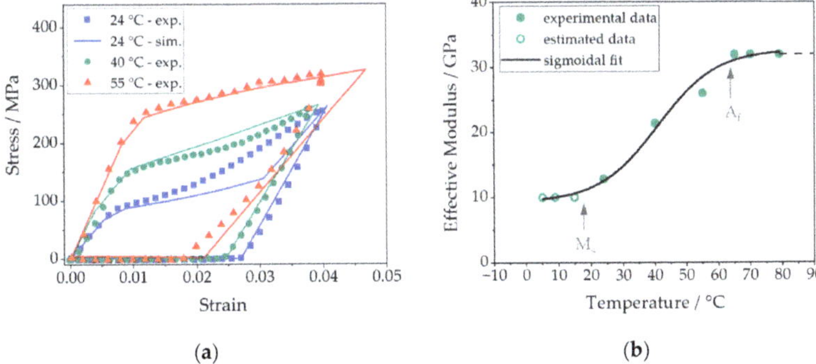

Figure 2. (a) Experimental and simulated stress–strain characteristics at different temperatures as indicated; (b) temperature-dependent effective modulus used in the present model to describe the mechanical response in the temperature regime of co-existing R- and A-phase.

2.2. Constitutive Modeling

The model used to describe the thermomechanical behavior of the SMA material has been adopted from Jaber et al. [21]. Contrary to other polycrystalline model approaches, it is based upon the strain tensor and temperature as control variables for the transformation process. In combination with the finite element method for finite strains and large displacements, this model has a very good convergence behavior. The change from a $(\sigma,\, T)$- to an equivalent $(\varepsilon,\, T)$-phase diagram is necessary to determine the transformation rates directly expressed by the strain rate. Skipping the stress calculation in the adopted algorithm reduces computation time. Similar to other phenomenological material models for SMAs, the present model consists of two laws: a thermomechanical and a kinetic law. Here, we present the general equations and refer the reader to references [21–23] for further details.

During the thermomechanical formulation, hypo-elastic plasticity modeling is conducted to describe the non-linear but reversible stress–strain behavior [24]. Here, the strain rate tensor is additively decomposed, as in the equation below,

$$\dot{\varepsilon} = \dot{\varepsilon}_{el} + \dot{\varepsilon}_{tr} \tag{1}$$

$\dot{\varepsilon}$ is the rate of the total strain, $\dot{\varepsilon}_{el}$ is the rate of the elastic strain, and $\dot{\varepsilon}_{tr}$ is the rate of the transformation strain. Considering the Jaumann rate of the Cauchy stress as the stress measure, the thermomechanical law is stated as below,

$$\dot{\sigma} = R \cdot \dot{\varepsilon}_{el} + k\,\dot{T} = R : \left(\dot{\varepsilon} - \dot{\varepsilon}_{tr}\right) + k\,\dot{T} \tag{2}$$

T is the current temperature, k is the thermal modulus tensor, and R is the fourth-order elastic modulus.

The martensite volume fraction ζ and transformation strain tensor ε_{tr} are selected as the internal state variables. These averaged quantities give a global account of the material behavior during thermomechanical loading. The kinetic law, therefore, describes the evolution of the internal state variables as below,

$$\dot{\varepsilon}_{tr} = \dot{\zeta} E, \tag{3}$$

where E is the transformation direction tensor that controls the orientation of created or recovered transformation strain for both forward and reverse transformations. E is defined as,

$$E = \varepsilon_{tr}^{max} \frac{\sigma'}{\|\sigma'\|} \text{ for } \dot{\zeta} > 0 \text{ (forward transformation)} \tag{4}$$

$$E = \varepsilon_{tr}^{max} \frac{\varepsilon_{tr}}{\|\varepsilon_{tr}\|} \text{ for } \dot{\varsigma} < 0 \text{ (reverse transformation)} \tag{5}$$

here, $\|\varepsilon\prime\| = \sqrt{\frac{3}{2(1+\nu)^2} \varepsilon\prime : \varepsilon\prime}$ indicates the tensor norm.

In forward transformation, the created strain follows the direction of the deviator σ', which controls shape change in the materials. In reverse transformation, the strain decrease follows the direction of the existing transformation strain ε_{tr}. With this constitutive relation, a tangent stiffness tensor is calculated, which relates local stress and strain increments. In the finite element implementation, an elastic predictor–transformation corrector return mapping algorithm is adopted. Geometric nonlinearity is treated with an updated Lagrangian formulation [25].

The phase diagram in Figure 3 has been obtained from evaluating the tensile loading data (see Table A1). In the model, we take a simplified view: In forward transformation, we assume the co-existence of R- and A-phase, which we combine to a single parent phase and describe by a temperature-dependent effective modulus determined from experimental material characterization. Above A_f, the stiffness of the unloaded material is constant, and we estimate the modulus of the starting phase to be 10 GPa below room temperature. Figure 2b depicts both the experimental and fitted effective modulus for different temperatures. The present model assumes the same mechanical behavior for tension and compression. In various NiTi alloys, a pronounced tension–compression asymmetry was found with typically higher transformation stress levels and lower transformation strain under compressive load [26]. An adaption of the model parameters will be possible in the future when experimental bending tests have been carried out.

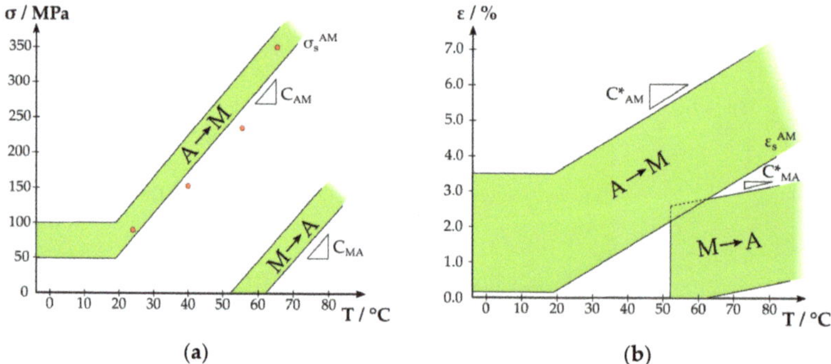

Figure 3. (**a**) Classical phase diagram obtained from experimental values (red dots); (**b**) conversion into modified phase diagram as adapted from [21].

2.3. Design and Engineering

The basic components of the origami-inspired microdevices are two triangular tiles that are interconnected by SMA folding microactuators. Each microactuator is designed as a double-beam structure as depicted in Figure 4a, which allows for resetting the memory shape by Joule heating and thereby localizing the heat to the folding region. While eliminating the need for an additional heating element, the double-beam structure also serves as a flexible active hinge connecting the individual tiles. The dimensions of the folding microactuator are mainly limited by the SMA foil thickness t_{SMA}. If ε_{max} denotes the maximum allowable tensile and compressive strain, which is located on the outer surfaces of the folding area, we can calculate the minimum admissible bending radius r as follows:

$$r > t_{SMA} \frac{1 - |\varepsilon_{max}|}{2 |\varepsilon_{max}|} \tag{6}$$

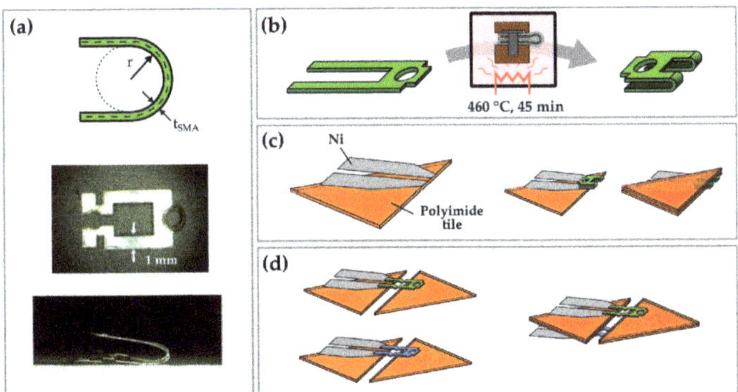

Figure 4. Fabrication and assembly of the folding microactuators: (**a**) microscope images of the micromachined NiTi foil before and after shape setting. (**b**) Schematic of laser micromachining and shape setting in a vacuum oven and (**c**) of electrical connection by gap welding and bonding to tiles. (**d**) Schematic of combining a protagonist and an antagonist microactuator for bi-directional folding.

For the given NiTi foil thickness of 20 µm and a strain limit of around 6%, the minimum bending radius is 157 µm. For our samples, we chose a bending radius of 250 µm to stay safely below the strain limit. We thereby obtained a minimum separation of the tiles of $\pi(r + t_{SMA}) = 0.82$ mm. The beam width was set to 1 mm and the beam separation to 2.5 mm. Additionally, two centering holes were designed, which are required later in the heat treatment procedure.

2.4. Fabrication

The manufacturing process of the folding microactuators is shown in Figure 4b–d). First, the NiTi foil is structured by laser cutting. The obtained structure is covered with titanium foils to prevent oxidation during the annealing step. By means of a pin and the centering holes, we can bring the stack of Ti and NiTi foils into the desired fold of 0° and a folding diameter of 500 µm. Then, shape setting is performed by heat treatment in a vacuum furnace at 460 °C for 45 min. In the next step (Figure 4c), the NiTi microactuators have to be attached to the polyimide tiles, which is performed here by adhesive bonding. Two nickel sheets are used as electrical connections, which are welded to the legs of the NiTi structure. The second tile is finally bonded to the NiTi structure to obtain the base element of the folding microactuator. Combining a protagonist microactuator with memory shape at 0° with an antagonist with memory shape at 360° (Figure 4d) enables bi-directional folding.

3. Characterization of Folding Actuation

The SMA folding actuator deforms quasi-plastically upon loading due to intermartensitic transformation. As shown in Figure 5, the shape-set folding microactuator has been mechanically unfolded beyond 360° at room temperature to achieve a visible deformation of approximately 210°. If the structure is then placed on a hot plate, the memory shape can be recovered by heating above A_f temperature. On subsequent cooling, the structure opens up again to about 45° due to internal stress induced by the large bending angle. Repeatable performance is achieved after five training cycles. However, no systematic investigation has been made into the fatigue life.

3.1. Uni-Directional Self-Folding

In the following, we present the results of the folding performance of a single SMA microactuator, which is quasi-plastically deformed to 180° and subsequently restored to its memory shape by Joule heating. The motion and temperature of the microactuator are tracked by an infrared camera. Different load cases are considered, ranging from free recovery, whereby no tiles are used, to the cases of Kapton and ceramic tiles, generating different gravitational forces. Figure 6 depicts the corresponding folding characteristics as a function of heating power, which correlates with the average temperature of the microactuator. For increasing load, the critical power to start the transformation shifts from about 150 to 500 mW. In all cases, full transformation is reached at a heating power of 600 mW. The corresponding critical temperatures increase from about 32 to 48 °C for the ceramic tile, which is due to the stress-induced increase of phase transformation temperatures. At the same time, the folding angle drops more steeply. Upon cooling, the memory shape is recovered completely due to the gravitational load of the ceramic tile.

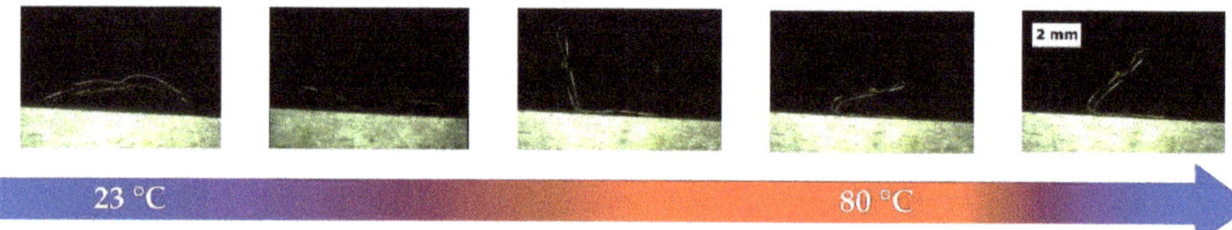

Figure 5. Folding sequence of a mechanically deformed SMA folding microactuator at room temperature. The memory shape can be recovered by heating on a hot plate above the austenite finish temperature.

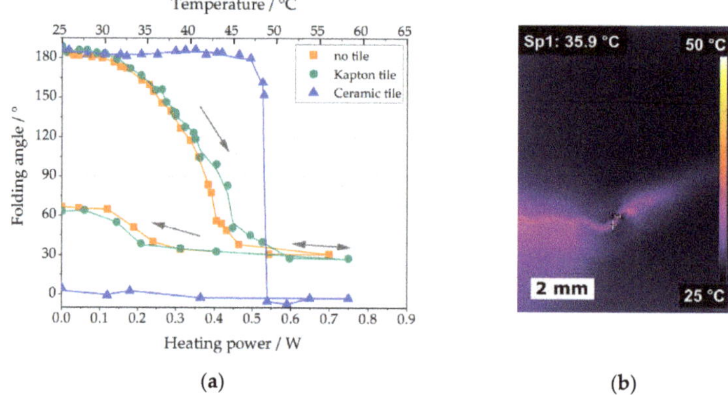

(a)   (b)

Figure 6. (a) Folding angles as a function of heating power and average temperature of the self-folding microactuator determined by camera tracking; (b) IR image taken during shape recovery.

These measurements prove the concept of folding and the tunability of the angular range by introducing a suitable additional load. However, at a small load, incomplete shape recovery is observed as a consequence of residual plastic deformation. Additional experiments prove that samples with an increased folding radius of 0.5 and 1 mm show less residual strain and, thus, a smaller difference between the folding angles in a hot and cold state. On the other hand, shape setting results in smaller folding angles in this case, as higher bending radii are associated with smaller strain during mechanical unfolding, causing reduced shape recovery [27].

In addition, we investigate the torque of the SMA microactuator through the blocking force that is generated during shape recovery. For this purpose, the SMA microactuator is positioned in two configurations (90° and 180°) below a load cell with one end fixed onto a substrate (see Figure 7a). The shape recovery during heating is blocked by the load cell, and the acting force is measured. The torque is calculated by multiplying the measured blocking force of the load cell with the distance L with respect to the fixation. Figure 7b shows the heating and cooling characteristics for each configuration. The observed hysteresis is attributed to the difference between the forward and reverse phase transformation temperatures. The maximum generated torque is determined to be 0.085 and 0.11 Nmm for the folding angles of 90° and 180°, respectively.

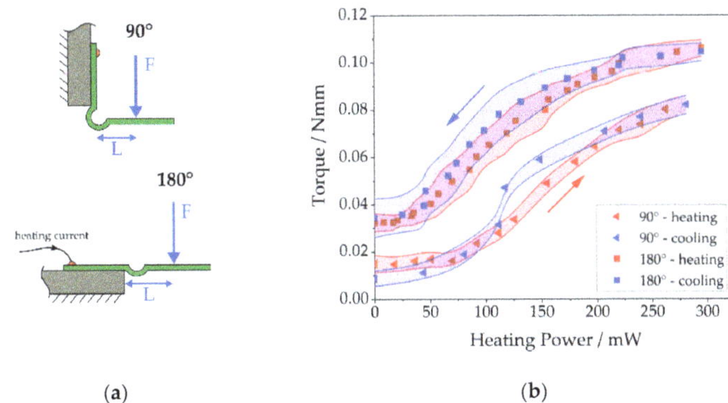

Figure 7. Measurement of blocking force versus folding angle of a single self-folding microactuator: (**a**) Schematic of experimental setup for the case of 90° and 180° folding angle; the resulting torque is obtained by the product of measured force F and distance L with respect to the fixation; (**b**) torque versus heating power characteristics upon heating and cooling.

3.2. Bi-Directional Self-Folding

By combining two counteracting microactuators, a protagonist and an antagonist, bi-directional self-folding can be achieved. Thereby, selective heating of only one microactuator at a given time is mandatory to avoid plastic deformation in the austenitic state. While the active actuator is heated, the antagonist in a cold state will be deformed. First, we determine the torque-folding angle characteristics of a single SMA microactuator at room temperature and under Joule heating. As in the previous experiment on single microactuators, the load cell is positioned above the actuator (Figure 8a), and the fixation of the tile is gradually rotated to adjust different folding angles. Using digital image processing, we evaluate the folding angle and the distance from the pin to fixation in order to derive the corresponding torque to the recorded blocking force. Figure 8b shows torque-folding angle characteristics for different average temperatures of the microactuators. The experimental setup allows for limited accuracy at small folding angles. In the initial elastic range, the stiffness increases for increasing average temperature. Joule heating up to 60 °C using a heating power of 110 mW causes only partial phase transformation. In this case, stress-induced formation of martensite causes a characteristic change in slope above a critical torque of about 0.55 Nmm and a plateau-like course of torque. Further heating to 80 °C results in full transformation to austenite, which is reflected by the quasi-linear loading characteristic. Yet, the folding microactuator is deformed plastically upon loading. Therefore, residual strain occurs upon unloading, which cannot be recovered.

The FE model is applied to the bending of a single self-folding microactuator to understand the folding motion and examine the torque in the range of small folding angles. For the 3D structure (Figure 9a), a zero-displacement boundary condition is applied to fix the two legs (green) to the tile below. For the blue part connecting the two beams to the second tile, a rigid body constraint is used, and unfolding is simulated by linearly increasing the moment on this part. Figure 9b depicts two different configurations (180° and 270°) and the corresponding volume distribution of martensite phase fraction at a homogeneous temperature of 24 °C. We can observe that the phase transformation does not occur uniformly along the folding axis but starts on the outer surfaces of the microactuator. For a folding angle of 270°, the maximum von Mises stress is 217 MPa and the maximum principal strain is 4.3%.

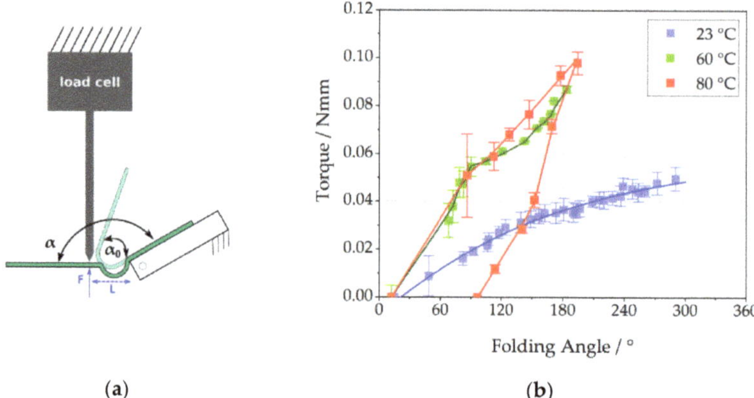

(a)   (b)

Figure 8. Measurement of blocking force versus folding angle of a single self-folding microactuator at different temperatures: (**a**) schematic of experimental setup; (**b**) torque-folding angle characteristics for different average temperatures of the microactuator.

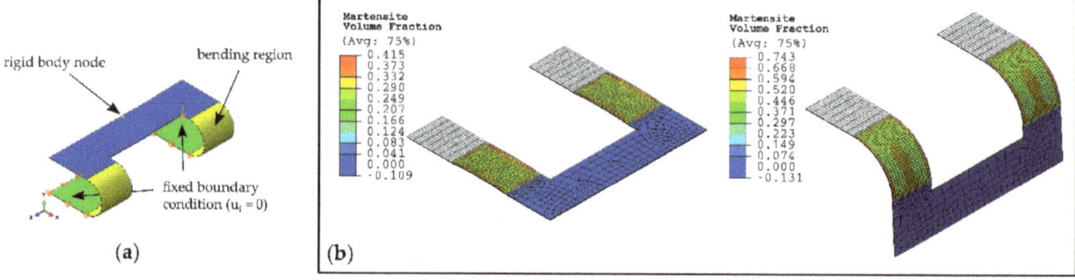

(a)   (b)

Figure 9. Finite element simulation of moment-controlled unfolding of a single microactuator at 24 °C: (**a**) boundary conditions of the 3D structure; (**b**) two folding configurations and corresponding local profile of martensite phase fraction indicated in color.

Simulations of torque-folding angle characteristics are carried out for different homogeneous temperatures, as shown in Figure 10a. Close to 160°, a kink occurs, which indicates a buckling effect that becomes increasingly pronounced for increasing temperature when the structure becomes stiffer. Figure 10b depicts the corresponding changes in martensite volume fraction in the bending region, covering the section between the fixation of the tiles and the connection of the two beams, see Figure 9. At 80 °C, the microactuator remains in an austenitic state, even for folding angles up to 180°. For larger deformation loads, stress-induced formation of oriented martensite occurs, which is fully reset upon unloading. At lower temperatures, stress-induced transformation to oriented martensite is initiated at lower folding angles between 90° and 180°. Thus, higher fractions of martensite occur for a given folding angle in this range. Below about 60 °C, oriented martensite can no longer be fully reset and, thus, folding angles are no longer recovered upon unloading. An animation of unfolding as well as the corresponding evolution of martensite phase fraction and von Mises stress for the cases of 24 and 80 °C can be found in videos S1–S4 of the Supplementary Materials, respectively.

Based on the torque-folding angle characteristics, the motion range of antagonistic self-folding microactuators can be predicted. Figure 11 shows torque versus folding angle of bi-directional microactuation for the case of 180° pre-deflection of protagonist and antagonist, thus a total of 360° with respect to each other. Zero position corresponds to a flat state (Position 1). The intersections of the characteristics of protagonist and antagonist correspond to the equilibrium positions obtained when selectively heating one microactuator at a given time. When the protagonist is heated exclusively, the folding angle α changes to Position 2 at $-100°$, while selective heating of the antagonist results in a folding angle at $+100°$ (Position 3). Thus, the equilibrium positions mark the maximum possible range of folding angles that could be achieved by antagonistic actuation. In an ideal case, bi-directional actuation allows covering the range of $-100° < \alpha < 100°$. The experimental results on real antagonistic devices, however, reveal lower folding angles due to additional forces in the devices depending on the details of fabrication. Figure 11 shows an example of bi-directional self-folding microactuation by infrared microcopy, showing reversible folding between $-44°$ and $+40°$. The folding motion during heating and cooling of the protagonist and antagonist can be found in videos S5 and S6 of the Supplementary Materials, respectively.

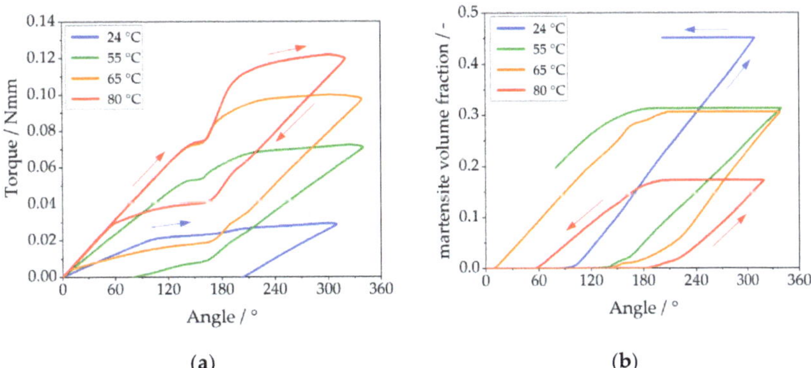

Figure 10. (**a**) FEM simulation of torque versus folding angle of a single microactuator for different homogeneous temperatures as indicated; (**b**) corresponding local martensite volume fraction versus folding angle in the bending region.

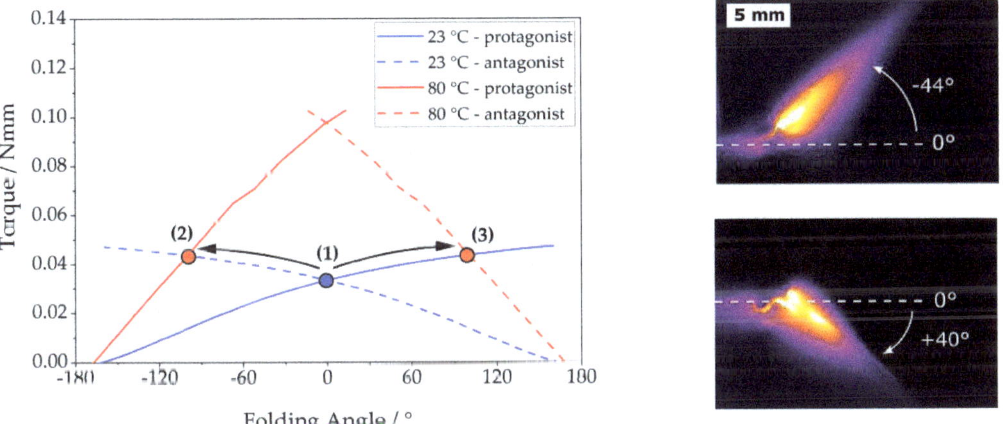

Figure 11. Torque versus folding angle characteristics of the protagonist and antagonist of an antagonistic microactuator for the case of pre-deflection by 360°. By heating either the protagonist or antagonist, the equilibrium point moves from (1) to either (2) or (3).

4. Conclusions

We present the design, fabrication, and characterization of uni- and bi-directional origami-inspired self-folding microactuators based on micromachined SMA foils. The basic design consists of two triangular tiles that are interconnected by double-beam-shaped SMA microactuators, forming active hinges that are controlled by Joule heating. A macromodel describing the engineering stress–strain characteristics of the SMA foil is set up and implemented in an FEM tool to simulate the torque and bending angles of corresponding SMA microactuators in the temperature range of actuation. For fabrication, we follow a rapid prototyping approach using laser cutting, separate heat treatment, and hybrid integration. Upon further miniaturization, these process steps will be replaced by a monolithic fabrication scheme. Characterization concentrates on uni-directional as well as bi-directional self-folding. Experiments on folding angles versus heating power on single microactuators demonstrate the self-folding and tunability of the angular range between $30°$ up to $180°$, depending on the gravitational force of the tiles. The finite element simulations qualitatively describe the main features of the observed torque-folding angle characteristics including the temperature dependence of stiffness, critical stress for intermartensitic transformation, and stress-induced formation of martensite. Based on this information, the performance of antagonistic microactuators including maximum torque and bi-directional folding range is predicted. First demonstrator devices reveal bi-directional actuation in the range of $-44°$ and $+40°$. This range remains well below the predicted limit of $\pm100°$, which indicates that there is considerable room for design improvements. Possible design extensions could include cascaded designs of folding microactuators and, e.g., additional magnetic forces to support self-folding. Further downscaling and increasing the number of tiles will enable a variety of functional shapes and motions based on programmable folding patterns. Possible applications are foreseen, e.g., in programmable adaptive microoptics and microfluidics.

Supplementary Materials: The following videos S1–S6 are available online at https://www.mdpi.com/article/10.3390/act10080181/s1. Video S1: Animation of unfolding and martensite phase fraction at 24 °C (24d-martensitefraction.mp4); Video S2: Animation of unfolding and von Mises stress in the bending region at 24 °C (24d-vonMises.mp4); Video S3: Animation of unfolding and martensite phase fraction at 80 °C (80g-martensitefraction.mp4); Video S4: Animation of unfolding and von Mises stress in the bending region at 80 °C (80g-vonMises.mp4); Video S5: Video of the folding motion during heating of the protagonist (protagonist-heating+cooling.mp4); Video S6: Video of the folding motion during heating of the antagonist (antagonist-heating+cooling.mp4).

Author Contributions: Conceptualization, M.K. and F.W., methodology, data curation, formal analysis, and visualization, L.S., simulations, G.K.T., writing—original draft preparation, L.S. and G.K.T., writing—review and editing, M.K. and F.W. All authors have read and agreed to the published version of the manuscript.

Funding: This research was funded by the German Research Foundation (Deutsche Forschungsgemeinschaft, DFG) within the priority program "Cooperative Multistage Multistable Microactuator Systems" (SPP2206).

Institutional Review Board Statement: Not applicable.

Data Availability Statement: The data presented in this study are available on request from the corresponding author. The data are not publicly available, because they are used in ongoing research.

Acknowledgments: We thank S. Wadhwa for performing the laser structuring of NiTi foils and H. Fornasier for UV lithography processing.

Conflicts of Interest: The authors declare no competing interest.

Appendix A

Table A1. List of parameters for finite element simulations.

Parameter	Abbreviation	Value
Martensite Start Temperature	M_s	19 °C
Martensite Finish Temperature	M_f	9 °C
Austenite Start Temperature	A_s	52 °C
Austenite Finish Temperature	A_f	62 °C
R-phase Peak Temperature	R_p	45 °C
Young's Modulus (starting phase)	E_A	12.9 GPa (at 24 °C) 21.4 GPa (at 40 °C) 24.3 GPa (at 55 °C) 32.5 GPa (at 80 °C)
Clausius Clapeyron coefficient	C_M	6.6 MPa/K
Maximum uniaxial transformation strain	ε_{tr}^{max}	0.035 (at 24 °C) 0.025 (40 °C, 55 °C, 80 °C)
Critical transformation start stress (A→M)	σ_{SCR}	~50 MPa
Poisson ratio	ν	0.33

References

1. Miskin, M.Z.; Dorsey, K.J.; Bircan, B.; Han, Y.; Muller, D.A.; McEuen, P.L.; Cohen, I. Graphene-based bimorphs for micron-sized, autonomous origami machines. *Proc. Natl. Acad. Sci. USA* **2018**, *115*, 466–470. [CrossRef]
2. Tolley, M.T.; Felton, S.M.; Miyashita, S.; Aukes, D.; Rus, D.; Wood, R.J. Self-folding origami: Shape memory composites activated by uniform heating. *Smart Mater. Struct.* **2014**, *23*, 94006. [CrossRef]
3. Jiang, M.; Gravish, N. Reconfigurable laminates enable multifunctional robotic building blocks. *Smart Mater. Struct.* **2021**, *30*. [CrossRef]
4. Cromvik, C.; Eriksson, K. *Airbag Folding Based on Origami Mathematics*; Chalmers University of Technology: Göteborg, Sweden, 2006.
5. Baek, S.-M.; Yim, S.; Chae, S.-H.; Lee, D.-Y.; Cho, K.-J. Ladybird beetle–inspired compliant origami. *Sci. Robot.* **2020**, *5*. [CrossRef]
6. Ho, M.; Kim, Y.; Cheng, S.S.; Gullapalli, R.; Desai, J.P. Design, development, and evaluation of an MRI-guided SMA spring-actuated neurosurgical robot. *Int. J. Rob. Res.* **2015**, *34*, 1147–1163. [CrossRef] [PubMed]
7. Momeni, F.; M.Mehdi Hassani.N, S.; Liu, X.; Ni, J. A review of 4D printing. *Mater. Des.* **2017**, *122*, 42–79. [CrossRef]
8. Peraza-Hernandez, E.A.; Hartl, D.J.; Malak, R.J., Jr.; Lagoudas, D.C. Origami-inspired active structures: A synthesis and review. *Smart Mater. Struct.* **2014**, *23*, 94001. [CrossRef]
9. Rogers, J.; Huang, Y.; Schmidt, O.G.; Gracias, D.H. Origami MEMS and NEMS. *MRS Bull.* **2016**, *41*, 123–129. [CrossRef]
10. Peraza-Hernandez, E.A.; Hartl, D.J.; Malak, R.J., Jr. Design and numerical analysis of an SMA mesh-based self-folding sheet. *Smart Mater. Struct.* **2013**, *22*, 94008. [CrossRef]
11. Shin, B.H.; Jang, T.; Ryu, D. J.; Kim, Y. A modular torsional actuator using shape memory alloy wires. *J. Intell. Mater. Syst. Struct.* **2016**, *27*, 1658–1665. [CrossRef]
12. Ho, M.; McMillan, A.; Simard, J.M.; Gullapalli, R.; Desai, J.P. Towards a Meso-Scale SMA-Actuated MRI-Compatible Neurosurgical Robot. *IEEE Trans. Robot.* **2011**, *2011*, 1–10. [CrossRef] [PubMed]
13. Boyvat, M.; Koh, J.-S.; Wood, R.J. Addressable wireless actuation for multijoint folding robots and devices. *Sci. Robot.* **2017**, *2*. [CrossRef]
14. Torres-Jara, E.; Gilpin, K.; Karges, J.; Wood, R.J.; Rus, D. Composable flexible small actuators built from thin shape memory alloy sheets. *IEEE Robot. Automat. Mag.* **2010**, *17*, 78–87. [CrossRef]
15. Hawkes, E.; An, B.; Benbernou, N.M.; Tanaka, H.; Kim, S.; Demaine, E.D.; Rus, D.; Wood, R.J. Programmable matter by folding. *Proc. Natl. Acad. Sci. USA* **2010**, *107*, 12441–12445. [CrossRef]
16. Paik, J.K.; Wood, R.J. A bidirectional shape memory alloy folding actuator. *Smart Mater. Struct.* **2012**, *21*, 65013. [CrossRef]
17. Jeong, J.W.; Yoo, Y.I.; Shin, D.K.; Lim, J.H.; Kim, K.W.; Lee, J.J. A novel tape spring hinge mechanism for quasi-static deployment of a satellite deployable using shape memory alloy. *Rev. Sci. Instrum.* **2014**, *85*, 25001. [CrossRef] [PubMed]
18. Otsuka, K. (Ed.) *Shape Memory Materials*; Cambridge Univ. Press: Cambridge, UK, 1999, ISBN 052144487X.
19. Miyazaki, S.; Otsuka, K. Deformation and transition behavior associated with the R-phase in Ti-Ni alloys. *MTA* **1986**, *17*, 53–63. [CrossRef]
20. Šittner, P.; Landa, M.; Lukáš, P.; Novák, V. R-phase transformation phenomena in thermomechanically loaded NiTi polycrystals. *Mech. Mater.* **2006**, *38*, 475–492. [CrossRef]
21. Jaber, M.B.; Smaoui, H.; Terriault, P. Finite element analysis of a shape memory alloy three-dimensional beam based on a finite strain description. *Smart Mater. Struct.* **2008**, *17*, 45005. [CrossRef]

22. Jaber, M.B. User Manual of the UMAT Subroutine for an SMA Strain Based Constitutive Model. Available online: www.researchgate.net/publication/280729999_User_Manual_of_the_UMAT_Subroutine_for_an_SMA_strain_based_constitutive_model (accessed on 27 July 2021).
23. Jaber, M.B.; Mehrez, S.; Ghazouani, O. A 1D constitutive model for shape memory alloy using strain and temperature as control variables and including martensite reorientation and asymmetric behaviors. *Smart Mater. Struct.* **2014**, *23*, 95026. [CrossRef]
24. ABAQUS. Abaqus Theory Manual V6.11: Mechanical Constitutive Models. *Dassault Systèmes Simulia Corp.* **2006**. Available online: 130.149.89.49:2080/v6.11/books/hhp/default.htm (accessed on 27 July 2021).
25. Bathe, K.-J. *Finite Element Procedures*; Prentice Hall: Englewood Cliffs, NJ, USA, 1996; ISBN 978-0133014587.
26. Gall, K.; Sehitoglu, H. The role of texture in tension–compression asymmetry in polycrystalline NiTi. *Int. J. Plast.* **1999**, *15*, 69–92. [CrossRef]
27. Seigner, L.; Bezsmertna, O.; Fähler, S.; Tshikwand, G.; Wendler, F.; Kohl, M. Origami-Inspired Shape Memory Folding Microactuator. In Proceedings of the 1st International Electronic Conference on Actuator Technology: Materials, Devices and Applications, Sciforum.net, 23–27 November 2020; MDPI: Basel, Switzerland, 2020; p. 8480.

 **actuators**

Article

Power Optimization of TiNiHf/Si Shape Memory Microactuators

Gowtham Arivanandhan [1], Zixiong Li [1], Sabrina M. Curtis [2,3], Lisa Hanke [2], Eckhard Quandt [2] and Manfred Kohl [1,*]

[1]   Institute of Microstructure Technology, Karlsruhe Institute of Technology (KIT), 76131 Karlsruhe, Germany
[2]   Institute for Materials Science, Kiel University (CAU), 24143 Kiel, Germany
[3]   Materials Science and Engineering, University of Maryland, College Park, MD 20742, USA
*   Correspondence: manfred.kohl@kit.edu

Abstract: We present a novel design approach for the power optimization of cantilever-based shape memory alloy (SMA)/Si bimorph microactuators as well as their microfabrication and in situ characterization. A major concern upon the miniaturization of SMA/Si bimorph microactuators in conventional double-beam cantilever designs is that direct Joule heating generates a large size-dependent temperature gradient along the length of the cantilevers, which significantly enhances the critical electrical power required to complete phase transformation. We demonstrate that this disadvantage can be mitigated by the finite element simulation-assisted design of additional folded beams in the perpendicular direction to the active cantilever beams, resulting in temperature homogenization. This approach is investigated for TiNiHf/Si microactuators with a film thickness ratio of 440 nm/2 µm, cantilever beam length of 75–100 µm and widths of 3–5 µm. Temperature-homogenized SMA/Si microactuators show a reduction in power consumption of up to 48% compared to the conventional double-beam cantilever design.

Keywords: shape memory films; TiNiHf films; bimorph shape memory microactuators; design optimization

Citation: Arivanandhan, G.; Li, Z.; Curtis, S.M.; Hanke, L.; Quandt, E.; Kohl, M. Power Optimization of TiNiHf/Si Shape Memory Microactuators. *Actuators* **2023**, *12*, 82. https://doi.org/10.3390/act12020082

Academic Editor: Wei Min Huang

Received: 13 January 2023
Revised: 11 February 2023
Accepted: 12 February 2023
Published: 15 February 2023

1. Introduction

Novel sensors and actuators have been developed in a wide range of MEMS applications to meet the ongoing trends of miniaturization and increases in functionality. Important fields of application are in silicon (Si) photonics requiring actuators for switching and tuning at small scales. The microactuators should allow for large strokes at small footprints and fabrication should be compatible with silicon technology. So far, the vast majority of actuators for nanophotonic switching and tuning is based on electrostatics, using either comb drive or gap-closing actuators [1,2]. Only in a few cases, other principles such as piezoelectric [3], thermal [4] or magnetic actuation [5] are applied.

Shape memory alloy (SMA) microactuators are an attractive solution owing to their distinct characteristics, such as a large actuation force, large displacement and low actuation voltage. The unique properties of SMA films and corresponding microstructures and nanostructures fabricated thereof provide a basis for the development of ultra-small actuators and systems with enhanced performance [6–8]. SMA/Si cantilever beam actuators have been demonstrated to show favorable scaling in the actuation performance with lateral dimensions down to 50 nm [9,10]. An ultra-compact nanophotonic SMA switch has been realized by the co-integration of SMA/Si nanoactuators and photonic waveguides on the same chip [11,12]. Cantilever beam actuators based on polymer/TiNiHf/Si composites enable bistable actuation that could be scaled from the mm-scale down to the nanoscale [13]. This performance is enabled by the large transformation hysteresis of TiNiHf films above room temperature.

The direct Joule heating of cantilever-based SMA bimorph actuators is known to generate a large size-dependent temperature gradient along the length of the cantilevers that can well exceed 50 °C/μm [10]. Because of this thermal scaling effect, the SMA layer undergoes phase transformation only in a local region, which significantly enhances power consumption to complete the phase transformation [11]. In the following, we present a novel approach to optimize the power consumption of cantilever-based TiNiHf/Si microactuators upon Joule heating through temperature homogenization. The electrical performance of temperature-homogenized microactuators is compared to non-optimized reference TiNiHf/Si microactuators.

2. TiNiHf Films

Recently, Curtis et al. demonstrated that the TiNiHf/Si and TiNiHf/SiO$_2$/Si film composites annealed at 635 °C showed favorable phase transformation properties including a phase transformation with a large thermal hysteresis above room temperature. This performance was shown to prevail upon downscaling the thickness of TiNiHf films down to 440 nm on Si substrates and 220 nm on SiO$_2$/Si substrates [13].

In this work, TiNiHf/Si cantilever beam microactuators are fabricated with the critical thickness of 440 nm using a DC magnetron multilayer sputter deposition approach as described in [13]. Amorphous TiNiHf films are magnetron sputtered on SOI substrates that have a 2 μm thick Si device layer and annealed at 635 °C for 5 min. The average film composition of sputtered films is measured using energy-dispersive x-ray spectroscopy and found to be Ti$_{40.4}$Ni$_{48}$Hf$_{11.6}$. Further details on the sputtering process and material's properties of the TiNiHf films can be found in [13]. Four-point electrical resistance measurements are carried out inside a cryostat to investigate the phase transformation of TiNiHf films constrained by a silicon-on-insulator (SOI) substrate. Quasi-stationary conditions are established by ramping the temperature step-wise and providing for sufficient waiting times in each step to guarantee that the influence of temperature change is negligible during measurement. Phase transformation temperatures are determined using the tangential method. Figure 1 shows the temperature-dependent electrical resistance of a TiNiHf film of 440 nm thickness upon cyclic heating and cooling.

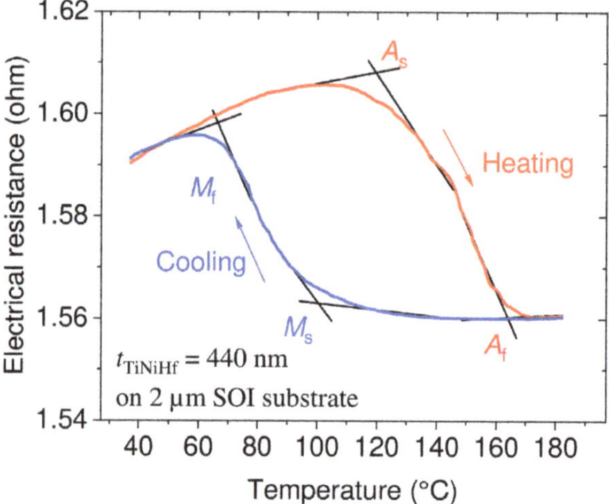

Figure 1. Temperature-dependent electrical resistance characteristic of a 440 nm thick TiNiHf film on a 2 μm thick Si device layer of a SOI substrate. The start and finish austenitic and martensitic transformation temperatures (As = 119.3 °C, Af = 163.7 °C, Ms = 100.2 °C, Mf = 65.6 °C) are determined using the tangential method.

The cyclic heating and cooling of the material reveal the typical martensitic phase transformation temperatures in line with the transformation temperatures and thermal hysteresis values reported for sputtered freestanding TiNiHf films [13]. The complete hysteresis loop shows the full phase transformation of the TiNiHf material. The characteristic phase transformation temperatures are determined using the tangential method. The hysteresis width is determined via the difference of average transformation temperatures $\overline{T}_A - \overline{T}_M$ is 67 °C ($\overline{T}_A = 1/2\left(A_s + A_f\right), = 1/2\left(M_s + M_f\right)$).

3. Modelling and Design Approach

The shape memory effect is described using a 1D Tanaka-type model that has been extended to isotropic SMA materials of a 3D shape [14,15]. A variable for martensite phase fraction $\xi_m(\sigma, T)$ is used that approximates the kinetics of the phase transformation by an exponential step function depending on the temperature and stress. In the phase transformation regime between austenitic and martensitic states, the material properties are described using a rule of mixture [16]. For instance, the thermal expansion coefficient of TiNiHf is approximated as a mixture of thermal expansion coefficients of martensite α_M and austenite α_A:

$$\alpha\left(T\right) = (1 - \xi_M(T))\cdot\alpha_A + \xi_M(T)\cdot\alpha_M \tag{1}$$

For the simulation of temperature profiles, the stationary heat equation for the TiNiHf/Si cantilever beam is solved numerically by approximating the temperature-dependent heat conductivity of the two layers. The heat conductivity of TiNiHf $\kappa_{TiNiHf}(T)$ is determined using the Wiedemann–Franz law [17] describing the heat conductivity of metals as a function of electrical conductivity σ:

$$\kappa_{TiNiHf}(T) = L_0\cdot\sigma_{TiNiHf}(T)\cdot T, \tag{2}$$

with the proportionality constant $L_0 = 2.45 \times 10^{-8}\frac{W\Omega}{K^2}$ [17]. The thermal conductivity of Si [18] is approximated using

$$\kappa_{Si}(T) = \frac{5.0105 \times 10^4 \frac{W}{m}}{T} - 19.24 \frac{W}{mK}, \tag{3}$$

Finite element simulations of the Joule heating of TiNiHf/Si cantilever actuators are performed by using COMSOL Multiphysics 6.0, which allows us to couple the models for the electrical current profiles and heat transfer. The resistive heating via electro-thermal coupling acts as a thermal stress on the solid–mechanics interface, causing differential thermal expansion as well as strain change due to martensitic phase transformation. The material properties of TiNiHf films are taken from a recent experimental study (See Table 1) [13]. The stress and electrical conductivity in the phase transformation regime are approximated using the rule of mixture as described before.

Table 1. Material properties for coupled electro-thermo-mechanical simulation of temperature profiles for TiNiHf/Si bimorph microactuators. The Poisson ratio of Si is included in the COMSOL material library. A polynomial function approximates the thermal expansion coefficient of Si [18]: $\alpha_{Si}(T) = (-3.0451 + 0.035705\cdot T - 7.98110^{-5}\cdot T^2 + 9.578310^{-8}\,T^3 - 5.891910^{-11}\cdot T^4 + 1.461410^{-14}\cdot T^5)10^{-6}\,1/K$.

	Si [19,20]	TiNiHf [13,21]
Electrical conductivity β_{SMA} (S/m)	10	$\sigma_{NiTiHf}(T)$
Thermal expansion coefficient α (1/K)	$\alpha_{Si}(T)$	$\alpha_A = 30 \times 10^{-6}$ $\alpha_M = 15 \times 10^{-6}$
Young's modulus E (GPa)	169	$E_m = 35$ $E_a = 83$
Poisson's ratio ν	0.22	0.39

A double-beam cantilever design with two contact pads is chosen for the TiNiHf/Si microactuators, as it facilitates Joule heating. A major concern with direct Joule heating is the scaling effects of heat transfer. Due to rapid heat transfer at the onset of the cantilever beams, Joule heating results in a temperature gradient along the beams, showing the highest values at the beam tip and lowest values close to substrate at the beam onset. As a consequence of this in-homogenous temperature profile, the phase fraction of martensite varies along the beam; thus, all material properties depending on the phase fraction vary along the beam as well. In particular, some sections of the beam may have already transformed to austenite, while other sections may still be in a martensitic state. Consequently, a large electrical power is required to heat the entire beam above the phase transformation temperature of about 164 °C, and actuation stroke is reduced.

Figure 2 shows the simulated temperature profiles along the x-direction of a TiNiHf/Si cantilever beam for various values of electrical power. At 15 mW of electrical power, the reference design exhibits a maximum tip temperature of 247 °C, while the beam onset is near room temperature (Figure 2a,c). Consequently, only about 50% of SMA material has reached the austenitic finish temperature (A_f = 164 °C) and the remaining 50% remains in the martensitic state. Further increasing the electric power accumulates a large amount of heat at the cantilever tip and, thereby, increases the temperature gradient considerably. At 21.4 mW, the reference TiNiHf/Si design already reaches a temperature gradient of 8 °C/μm.

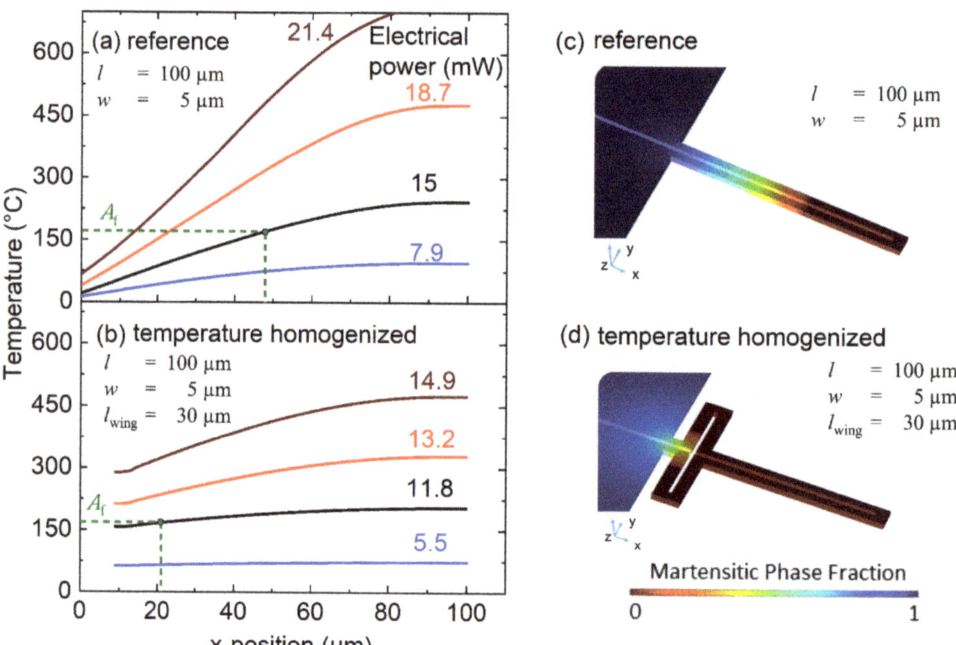

Figure 2. (**a,b**): Coupled finite element simulation of the temperature profiles along x-direction of a TiNiHf/Si cantilever beam upon Joule heating for various values of electrical power. The beam length *l*, beam width *w* and length of folded beams *l*$_{wing}$ are indicated. The reference design without additional folded beams shows much larger temperature gradients compared to the temperature-homogenized design with additional folded beams; (**c,d**): contour plots of martensite phase fraction for reference and temperature-homogenized design at an electrical power of 15 mW and 11.8 mW, respectively.

This disadvantage can be mitigated by the design of additional folded beams with a direction perpendicular to the active cantilever beams, giving rise to much more homogeneous temperature profiles, as shown in Figure 2b,d. Already at the electrical power of 11.8 mW, the maximum tip temperature reaches 210 °C, while only 22% of SMA material remains in a martensitic state. The maximum power required to complete the phase transformation is about 12.25 mW. Thus, the presented approach of temperature homogenization largely eliminates the issue of different phase states at local regions of the cantilever beam.

4. Fabrication of TiNiHf/Si Bimorph Microactuators

Figure 3 shows a schematic of the process flow for fabrication of the TiNiHf/Si microactuators. This process follows our previous approach of micromachining and nanomachining a SOI wafer prior to SMA film deposition. It has the advantage that the SMA's functionality is not impaired by silicon technology and film delamination is prevented [22]. The starting substrate is a silicon-on-insulator (SOI) chip with a 2 µm device layer and 2 µm buried oxide layer. In the first steps, the pattern of cantilever beams is transferred to a chromium layer by using e-beam lithography (EBL) (Figure 3a,b) and a lift-off technique (Figure 3c). A chromium layer of 30 nm is used as a hard mask to etch the Si device layer via cryogenic reactive ion etching (RIE) (Figure 3d). In the next step, the chromium layer is removed, and the buried oxide layer is selectively removed via wet etching using hydrofluoric HF acid to obtain freestanding Si beam structures (Figure 3e). In the final step, the TiNiHf film is deposited via DC magnetron sputtering near room temperature to obtain freestanding TiNiHf/Si bimorph microstructures. Amorphous SMA films are then crystallized at 635 °C via rapid thermal annealing. Scanning electron micrographs of the fabricated TiNiHf/Si microactuators are shown in Figure 4. The TiNiHf layer of 440 µm and Si layers of 2 µm thickness can be distinguished clearly.

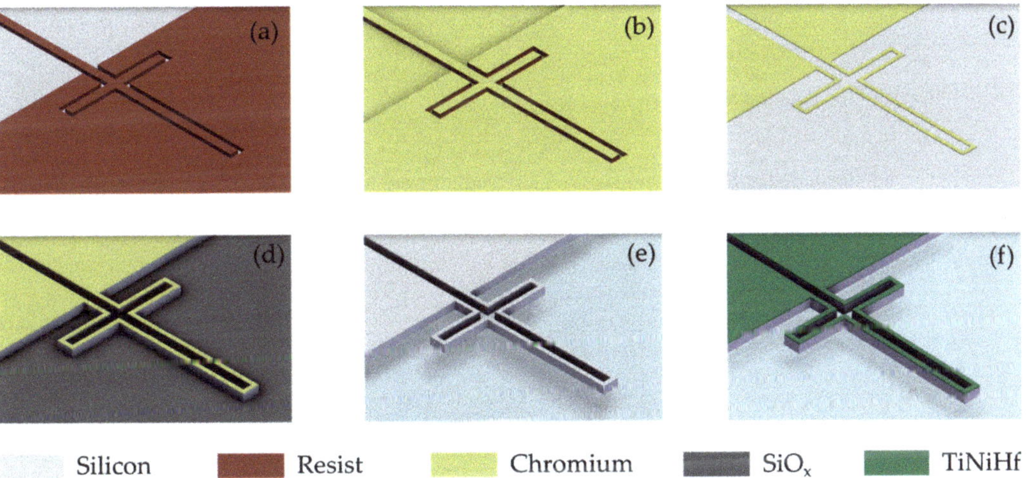

Figure 3. Schematic process flow for fabrication of TiNiHf/Si microactuators using an SOI wafer with 2 µm Si device layer, (**a**) patterning of resist using EBL, (**b**) chromium deposition, (**c**) lift-off of chromium, (**d**) RIE of Si using cryogenic etching, (**e**) stripping of chromium and selective etching of SiOx sacrificial layer and (**f**) sputtering and annealing of TiNiHf thin film.

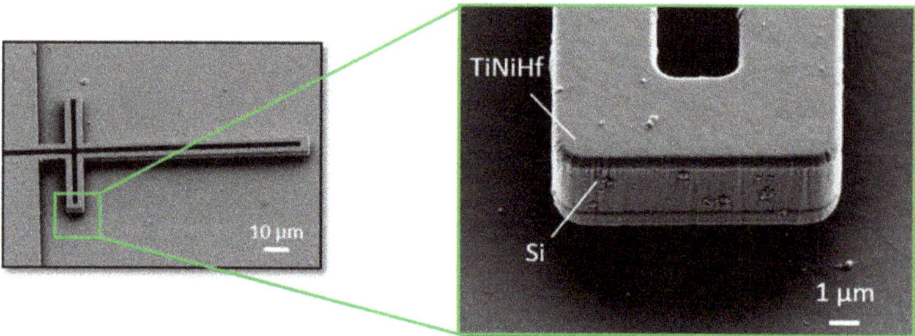

Figure 4. Scanning electron micrographs of a freestanding TiNiHf /Si microactuator with folded beam structure for temperature homogenization. The thicknesses of the TiNiHf film and Si layer are 440 nm and 2 μm, respectively. Dimensions of cantilever beams: 75 μm length (l), 3 μm width (w) and 30 μm length of folded beam structure (l_{wing}).

5. Evaluation of Critical Electrical Power

Electrical resistance measurements are performed as a function of electrical power in situ in a scanning electron microscope (SEM) using the four-point method under stationary equilibrium conditions as described in Section 2. Nanomanipulators are used to establish electrical interconnections to individual microactuators as illustrated in Figure A1 in Appendix A. Figure 5 shows typical electrical resistance characteristics for different TiNiHf/Si microactuators without folded beam structures (reference design) and with folded beam structures for temperature homogenization. The electrical resistance characteristics show a complete hysteresis loop upon Joule heating and cooling, revealing the required electrical power for completing the phase transformation. Upon Joule heating, the average electrical resistance shows a characteristic drop for increasing electrical power, reflecting the course of the phase transformation. In the reference design, the resistance drop occurs gradually, as the phase transformation progresses gradually along the cantilever beam from the tip to the onset. In contrast, the temperature-homogenized design exhibits a much sharper drop in electrical resistance due to a larger volume fraction of SMA material transforming at a given heating power. In the present case (Figure 5a), the critical heating power required for a complete hysteresis loop reduces from 15 mW to about 11.8 mW corresponding to a reduction in power consumption by 21.8%. Figure 5b shows a comparison of TiNiHf /Si microactuators with different lengths of folded beam structures l_{wing} to illustrate their effect on temperature-homogenization. When increasing l_{wing} from 30 to 40 μm, the heating power required for complete phase transformation reduces from 7.9 to 5.5 mW, while the corresponding reference design requires about 10 mW. Thus, the reduction in power consumption reaches 45% in the latter case.

Table 2 summarizes the results concerning the critical electrical power required for a complete hysteresis loop for each sample. By narrowing the width of the cantilever beams from 5 to 3 μm (compare samples Ref100, #1 and #2), an additional power saving from 21.8% to 48% is achieved. The reduced cross-section hinders heat transfer at the cantilever beam onset and, thus, favors temperature homogenization and power optimization. However, reducing the beam width for power optimization is limited largely by requirements on the mechanical performance of the microactuators. Increasing the length of the folded beams from 30 μm to 40 μm (compare samples Ref75, #3 and #4) increases the temperature-homogenization effect and enhances the power saving from 21% to 45%. A further increase in the folded beam length may further improve power saving, but might lead to mechanical instability. In addition, design limitations due to fabrication constraints have to be considered in this case.

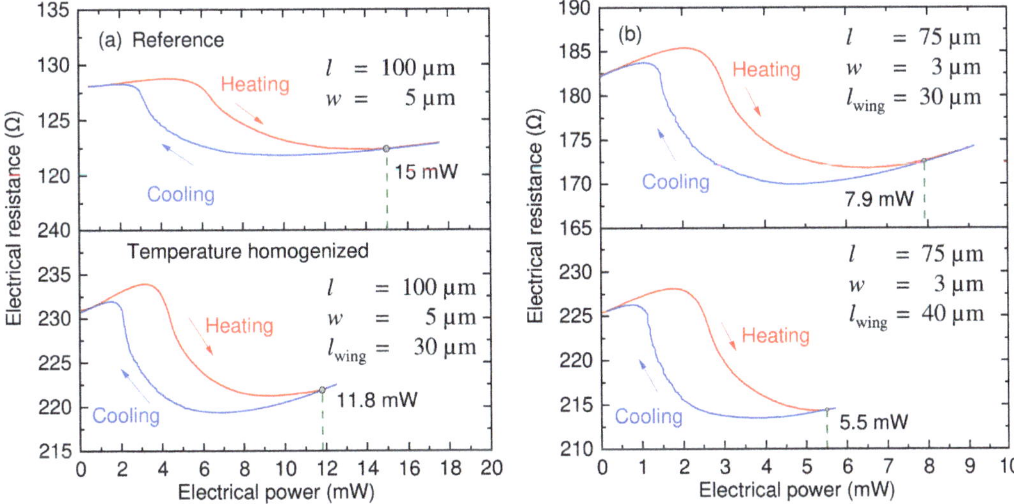

Figure 5. In situ SEM measurement of electrical resistance as a function of electrical heating power for different TiNiHf/Si microactuators. The beam length l, beam width w and length of folded beams l_{wing} are indicated. The required electrical power to complete the phase transformation is highlighted. (**a**) Comparison of reference (top) and temperature-homogenized design (bottom), (**b**) temperature-homogenized design for different lengths of the folded beams (l_{wing}) of 30 and 40 µm.

Table 2. Summary of experimental results of critical electrical power upon Joule heating required for a complete hysteresis loop and corresponding power saving for different geometries of TiNiHf/Si microactuators.

Geometry/Parameter	Ref100	#1	#2	Ref75	#3	#4
Beam length (l), µm	100	100	100	75	75	75
Beam width (w), µm	5	5	3	3	3	3
Folded beam length [a] (l_{wing}), µm	0	30	30	0	30	40
Critical electrical power (E_c), mW	15	11.8	7.8	10	7.9	5.5
Power saving,% (as compared to reference)		21.8	48		21	45

[a] The wing width always coincides with the beam width of the cantilever.

The temperature profile along the beam cannot be measured accurately due to size limitations. Therefore, we estimate the maximum temperatures at the cantilever tip by our simulations of Joule heating-induced temperature profiles. Figure 6 shows simulated maximum tip temperatures for TiNiHf/Si microactuators with two different beam lengths and widths. For instance, temperature-homogenized TiNiHf/Si microactuators with dimensions $l = 75$ µm, $w = 3$ µm and lwing = 30 µm show a maximum tip temperature of 212 °C at a power of 8 mW, while the reference microactuators show a maximum tip temperature of 230 °C at 10 mW. These results indicate that a lower heating power is required to reach a certain maximum tip temperature for the temperature homogenized design.

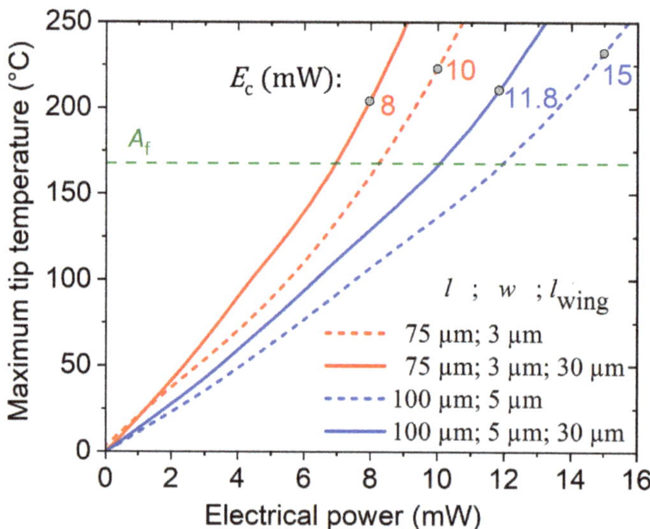

Figure 6. Simulated characteristics of maximum temperature at the cantilever beam tip as a function of electrical heating power. Measured values of electrical power (dots) are compared to simulation results to estimate the maximum temperature at the cantilever beam tip.

6. Conclusions

In this work, we address the size-dependent effect of large temperature gradients upon direct Joule heating. We introduce a novel method to homogenize the temperature in cantilever-based SMA/Si microactuators by designing additional folded beams with a direction perpendicular to the active cantilever beams. This approach is evaluated for a series of TiNiHf/Si double-beam cantilevers that are fabricated by a combination of Si micromachining and magnetron sputter deposition of TiNiHf films. In situ electrical resistance measurements are performed to assess the critical electrical power required to achieve complete phase transformation for beam cantilevers with different beam lengths, widths, and lengths of additional folded beams. Our results reveal significant power savings of up to 48% compared to reference designs without temperature homogenization. We conclude that this approach of temperature homogenization largely eliminates the issue of inhomogeneous phase transformation along the beam cantilevers upon Joule heating. It discloses a simple measure for power optimization. In addition, the mechanical performance will be improved, as it also benefits from the larger fraction of SMA material undergoing the phase transformation.

Author Contributions: M.K. and E.Q. developed the concept; G.A. developed the simulation model, performed the simulations and experiments, analyzed the data and wrote the draft of the manuscript; Z.L. contributed to experiments and data analysis; S.M.C. and L.H. designed the TiNiHf experiments and analyzed the data; M.K. supervised the work, reviewed and edited the manuscript. All authors have read and agreed to the published version of the manuscript.

Funding: This research has received funding from the German Research Foundation (DFG) within the priority program "SPP2206—Cooperative Multistage Multistable Microactuator Systems".

Data Availability Statement: The data presented in this study is available upon request from the corresponding author. They are not publicly available as they are being used in ongoing research.

Acknowledgments: This work was partly carried out with the support of the Karlsruhe Nano Micro Facility (KNMF, www.knmf.kit.edu), a Helmholtz Research Infrastructure at Karlsruhe Institute of Technology (KIT, www.kit.edu).

Conflicts of Interest: The authors declare no conflict of interest.

Abbreviations

The following abbreviations are used in this manuscript:

EBL	Electron beam lithography
EDX	Energy-dispersive X-ray spectroscopy
FEM	Finite element modeling
MEMS	Micro-electro-mechanical systems
RIE	Reactive ion etching
SMA	Shape memory alloy
SOI	Silicon-on-insulator

Appendix A

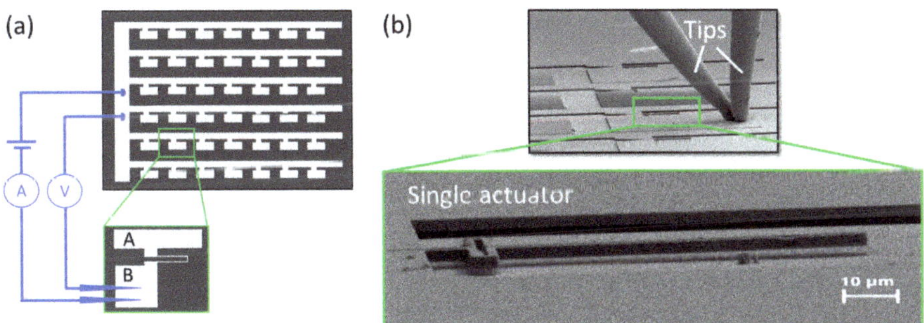

Figure A1. Experimental setup used for in situ SEM measurements of electrical resistance. (**a**) Schematic of the chip layout consisting of many microactuators arranged in an array and electrical interconnections. Each microactuator has two contact pads (A and B) for electrical interconnection, whereby the contact pads *A* of each microcactuator are electrically connected with each other and with two large pads being interconnected via wire bonding. In addition, two nanomanipulators are used to establish electrical interconnections to the second contact pads *B* of each single microactuator in a time sequence one after another. (**b**) SEM images taken during in situ electrical measurement showing a single microactuator and the nanomanipulator tips used for electrical interconnection.

References

1. Chollet, F. Devices based on co-integrated MEMS actuators and optical waveguide: A review. *Micromachines* **2016**, *7*, 18 [CrossRef] [PubMed]
2. Du, H.; Chau, F.; Zhou, G. Mechanically-tunable photonic devices with on-chip integrated MEMS/NEMS actuators. *Micromachines* **2016**, *7*, 69. [CrossRef] [PubMed]
3. Masmanidis, S.; Karabalin, R.; De Vlaminck, I.; Borghs, G.; Freeman, M.; Roukes, M. Multifunctional nanomechanical systems via tunably coupled piezoelectric actuation. *Science* **2007**, *317*, 780. [CrossRef] [PubMed]
4. Guha, B.; Lipson, M. Controlling thermo-optic response in microresonators using bimaterial cantilevers. *Opt. Lett.* **2015**, *40*, 103. [CrossRef] [PubMed]
5. Kim, P.; Hauer, B.; Clark, T.; Sani, F.F.; Freeman, M.; Davis, J. Magnetic actuation and feedback cooling of a cavity optomechanical torque sensor. *Nat. Commun.* **2017**, *8*, 1355. [CrossRef] [PubMed]
6. Kohl, M. *Shape Memory Microactuators*; Springer: Berlin/Heidelberg, Germany, 2004.
7. Miyazaki, S.; Fu, Y.; Huang, W. (Eds.) *Thin Film Shape Memory Alloys: Fundamentals and Device Applications*; Cambridge University Press: Cambridge, UK, 2009.
8. Stachiv, I.; Alarcon, E.; Lamac, M. Shape memory alloys and polymers for MEMS/NEMS applications: Review on recent findings and challenges in design, preparation, and characterization. *Metals* **2016**, *11*, 415. [CrossRef]
9. Lay, C.; Aseguinolaza, I.; Chernenko, V.; Kohl, M. Development of ferromagnetic shape memory alloy–silicon bimorph nanoactuators. In Proceedings of the 13th IEEE Conference in Nanotechnology (IEEE-NANO), Beijing, China, 5–8 August 2013.
10. Kohl, M.; Schmitt, M.; Backen, A.; Schultz, L.; Krevet, B.; Fähler, S. Ni-Mn-Ga shape memory nanoactuation. *Appl. Phys. Lett.* **2014**, *104*, 4. [CrossRef]

11. Lambrecht, F.; Aseguinolaza, I.; Chernenko, V.; Kohl, M. Integrated SMA-based NEMS actuator for optical switching. In Proceedings of the 29th IEEE International Conference on Micro Electro Mechanical Systems (MEMS), Shanghai, China, 24–28 January 2016.
12. Rastjoo, S.; Fechner, R.; Bumke, L.; Kötz, M.; Quandt, E.; Kohl, M. Development and co-integration of a SMA/Si bimorph nanoactuator for Si photonic circuits. *Microelectron. Eng.* **2020**, *225*, 111257. [CrossRef]
13. Curtis, S.; Sielenkämper, M.; Arivanandhan, G.; Dengiz, D.; Li, Z.; Jetter, J.; Hanke, L.; Bumke, L.; Quandt, E.; Wulfinghoff, S.; et al. TiNiHf/SiO$_2$/Si shape memory film composites for bi-directional micro actuation. *Int. J. Smart Nano Mater.* **2022**, *13*, 293–314. [CrossRef]
14. Tanaka, K.; Nagaki, S. A thermomechanical description of materials with internal variables in the process of phase transitions. *Ing. Arch.* **1982**, *51*, 287–299. [CrossRef]
15. Tanaka, K. A thermomechanical sketch of shape memory effect: One-dimensional tensile behavior. *Res. Mech.* **1986**, *18*, 251–263.
16. Kohl, M.; Krevet, B. 3D Simulation of a Shape Memory Microactuator. *Mater. Trans.* **2002**, *43*, 1030–1036. [CrossRef]
17. Kopitzki, K.; Herzog, P. *Einführung in die Festkörperphysik*, 5th ed.; Teubner: Stuttgart, Germany, 2004.
18. Carruthers, J.A.; Geballe, T.H.; Rosenberg, H.M.; Ziman, J.M. The thermal conductivity of germanium and silicon between 2 and 300° K. *Math. Phys. Sci.* **1957**, *238*, 502–514.
19. Watanabe, H.; Yamada, N.; Okaji, M. Linear thermal expansion coefficient of silicon from 293 to 1000 K. *Int. J. Thermophys.* **2004**, *25*, 221–236. [CrossRef]
20. Jeong, J. Evaluation of elastic properties and temperature effects in Si thin films using an electrostatic microresonator. *J. Microelectromech. Syst.* **2003**, *12*, 524–530. [CrossRef]
21. Benafan, O. Deformation and Phase Transformation Processes in Polycrystalline NiTi and NiTiHf High Temperature Shape Memory Alloys. Ph.D. Thesis, University of Central Florida, Orlando, FL, USA, 2012.
22. Lay, C.; Aseguinolaza, I.; Chernenko, V.; Kohl, M. In-situ characterization of ferromagnetic shape memory alloy/silicon bimorph nanoactuators. In Proceedings of the 14th IEEE International Conference on Nanotechnology, Toronto, ON, Canada, 18–21 August 2014; pp. 192–195.

Article

Design and Optimal Control of a Multistable, Cooperative Microactuator

Michael Olbrich [1,*], Arwed Schütz [2], Tamara Bechtold [2] and Christoph Ament [1,*]

1   Chair of Control Engineering, University of Augsburg, Eichleitnerstraße 30, D-86159 Augsburg, Germany
2   Research Group for Modelling and Simulation of Mechatronic Systems, Jade University of Applied Sciences, D-26389 Wilhelmshaven, Germany; arwed.schuetz@jade-hs.de (A.S.); tamara.bechtold@jade-hs.de (T.B.)
*   Correspondence: michael.olbrich@uni-a.de (M.O.); christoph.ament@uni-a.de (C.A.)

Abstract: In order to satisfy the demand for the high functionality of future microdevices, research on new concepts for multistable microactuators with enlarged working ranges becomes increasingly important. A challenge for the design of such actuators lies in overcoming the mechanical connections of the moved object, which limit its deflection angle or traveling distance. Although numerous approaches have already been proposed to solve this issue, only a few have considered multiple asymptotically stable resting positions. In order to fill this gap, we present a microactuator that allows large vertical displacements of a freely moving permanent magnet on a millimeter-scale. Multiple stable equilibria are generated at predefined positions by superimposing permanent magnetic fields, thus removing the need for constant energy input. In order to achieve fast object movements with low solenoid currents, we apply a combination of piezoelectric and electromagnetic actuation, which work as cooperative manipulators. Optimal trajectory planning and flatness-based control ensure time- and energy-efficient motion while being able to compensate for disturbances. We demonstrate the advantage of the proposed actuator in terms of its expandability and show the effectiveness of the controller with regard to the initial state uncertainty.

Keywords: cooperative microactuator; multistability; magnetic levitation; feedback control

Citation: Olbrich, M.; Schütz, A.; Bechtold, T.; Ament, C. Design and Optimal Control of a Multistable, Cooperative Microactuator. *Actuators* **2021**, *10*, 183. https://doi.org/10.3390/act10080183

Academic Editor: Micky Rakotondrabe

Received: 12 July 2021
Accepted: 31 July 2021
Published: 4 August 2021

1. Introduction

Multistability is an important system property found in a variety of microactuators and corresponds to possessing multiple stable actuator positions. While it is most commonly used in switches with two stable resting positions, extending the number of equilibrium points opens up new possibilities in the functionality and application flexibility of microsystems. Especially for actuators with large working ranges, which are currently the focus of research, multistability may prove useful.

In order to enlarge the operating ranges of microactuators, mechanical connections resulting in large restoring forces have to be overcome. Therefore, a considerable number of actuator concepts have been proposed, which can be distinguished in particular on whether the entire actuator or a single object is moved. The former includes impact mechanisms, which generate a stepwise motion by repeated impact between an internal mass and a stopper and can be realized using different actuation principles [1–3]. Stick-slip actuators also move in small steps, but exploit different friction effects. Therefore, movements of both high precision [4] and large displacements [5] are possible. Progress has also been made in terms of new concepts, such as legged platforms, comparable to a linear motor [6,7]. In these approaches, the actuator is driven by traveling waves, generated by piezoelectric MEMS. Large travel ranges can also be achieved by more complex setups, such as the microrobot of Floyd et al. [8]. In these cases, there are no restoring forces driving the actuator back into its initial state. Thus, they are multistable since every position corresponds to a steady state when no external input is applied.

For many applications, however, the motion of a single passive object is sufficient. This typically results in less complex designs, simplifying the assembly process. In the simplest case, a small object, e.g., a magnet, can be used for a bistable switching application [9], or it is moved between these two stable positions using feedback control [10]. The advantage of the magnetic principle lies in the fact that both equilibrium positions are asymptotically stable in their vicinity, even without further input. Other typical applications include linear micro motors [11,12] and micro conveyor systems [13]. As proposed by Poletkin et al., conveyor systems can also be realized by arrays of inductive suspensions [14]. These have in common that they are based on levitation of the moved object, but need constant energy input to generate asymptotically stable equilibria. Another important criterion corresponds to the achieved degree of freedom (DoF). Without adaption, these actuators can only be used for planar movements and thus are strongly limited in the vertical direction.

In order to allow a flexible application in both the horizontal and vertical directions, we previously presented a magnetic actuator moving a cylindrical permanent magnet with one DoF [15]. To this end, finite element models (FEM) were used to calculate the magnetic forces, and their complexity was reduced by different approaches, i.e., a lookup-table, a semi-analytical compact model, and a parametric reduced order model [16,17]. These methods were then compared in terms of simulation time and accuracy, and their usefulness for a controller design was demonstrated. The actuator consists of permanent magnetic rings, generating two stable equilibrium positions: a lower one in contact with a piezoelectric stack actuator, and a second one located at a predefined height, where the magnetic object is stably levitated. Fast and efficient transient motion between these positions is handled by cooperation of the stack actuator, enabling a fast initial acceleration of the magnet to be moved, and a controlled catch at the second stable equilibrium, using electromagnets.

A major drawback lies in the merely bistable design and the restriction of the movable object to remain below the center of the electromagnetic solenoid, due to the limited controllability. In this work, we therefore aim to extend the proposed design by multiple solenoids and ring magnets, such that a greater number of resting positions is generated. In order to reduce electrical currents within the solenoids, a cooperative mechanism between the piezoelectric actuator and multiple solenoids is exploited. Moreover, an optimal control strategy is used to find a trade-off between low solenoid currents and time-optimal transitions. The effectiveness of the control scheme is demonstrated in the simulation, and the expandability of the actuator is discussed. The advantage of the proposed concept lies in its extensibility to larger working ranges as well as the generalizable control algorithm for an arbitrarily extended actuator.

2. Methods

The microactuator setup investigated in this work consists of a movable, permanent magnetic proof mass guided by a glass tube, a piezoelectric stack actuator, two permanent ring magnets and two solenoids. In contrast to our previous work [15], the second magnet is used to imprint a second levitating equilibrium position, rather than weakening the field of the first magnet. The additional solenoid allows a less restrictive motion range of the proof mass, which was an issue of the initial setup. The basic design is sketched in Figure 1. The grayed-out part illustrates its expandability, which is discussed Section 3.1.

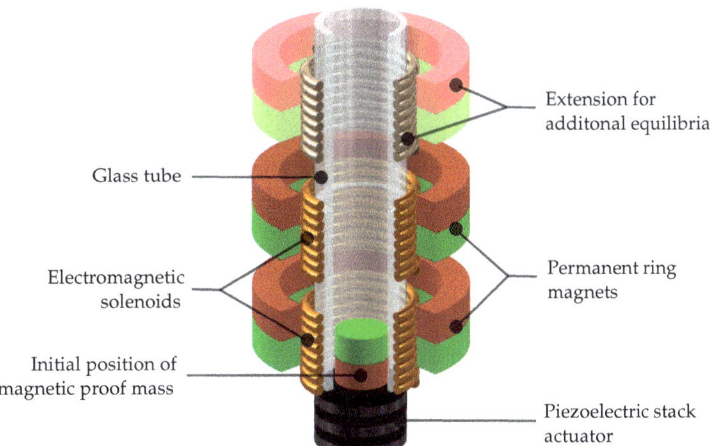

Figure 1. The multistable microactuator consists of a movable, cylindrical magnet within a guiding glass tube, a piezoelectric stack actuator (kick actuator), permanent magnets and electromagnets (catch actuators). The transparent part depicts a possible extension to a third levitating equilibrium.

The working principle is as follows: Initially, the proof mass is placed on top of the stack actuator, which corresponds to the lower resting position. In order to obtain an impulse-like acceleration, the inverse piezoelectric effect is used to kick the cylindrical magnet upwards. Subsequently, the electromagnetic solenoids are switched on and cooperatively catch it at the desired resting position in a controlled way. Since these positions are asymptotically stable by design, the proof mass is suspended without the need for further energy input. Concerning the transitions between the levitation equilibria, as well as the downwards motion, mere electromagnetic actuation is applied. The control objective is to achieve fast movements with minimum solenoid current. For this purpose, a mathematical system model containing all relevant effects is derived, and an optimal model-based control strategy is presented.

2.1. Modeling

The overall system can be described by the dynamics of each individual component and the respective couplings between them. The model is adapted from [15] and generalized for an arbitrary number of magnets and solenoids. Let us assume that the setup comprises m ring magnets and n electromagnets. The dynamic behavior of the current $i_k(t)$ of the k-th solenoid with inductance L_k, wire resistance R_k and input voltage $u_k(t)$ is then given by the following:

$$L_k \frac{\mathrm{d}i_k(t)}{\mathrm{d}t} = u_k(t) + u_{\mathrm{ind},k}(t) - R_k i_k(t) , \tag{1}$$

where $u_{\mathrm{ind},k}(t)$ is the voltage induced due to the change of electric current within the other solenoids. Given the coupling inductance $L_{j,k}$ describing the influence of the j-th solenoid on the k-th solenoid, the induced voltage is defined by the following.

$$u_{\mathrm{ind},k}(t) = -\sum_{j \neq k}^{n} L_{j,k} \frac{\mathrm{d}i_j(t)}{\mathrm{d}t} . \tag{2}$$

The current of each solenoid affects the proof mass with vertical position $z(t)$ in terms of an electromagnetic force $F_{\mathrm{em},k}(z, i_k)$, which is defined below in (9).

Similar to [18], we assume the piezoelectric actuator to behave like a linear second order system but neglect the hysteresis effect. Although the literature suggests more complex, nonlinear dynamics [19,20], this approximation seems justified, due to the short time of contact between the piezo and the proof mass. Therefore, we model the deformation $d(t)$ of the piezo as follows:

$$M\ddot{d}(t) = -Mg - c_A\dot{d}(t) - k_A d(t) - F_c(t) + F_u(t),\tag{3}$$

with the effective mass M, gravitational acceleration g, damping coefficient c_A and stiffness k_A. The force that results in the deformation by applying a voltage $u_A(t)$ is considered by $F_u(t)$ and the interaction with the proof mass corresponds to the contact force $F_c(t)$. In order to describe $F_u(t)$ in dependence of the applied voltage, we consider the static piezoelectric behavior. As proposed in [21], we use a linear relation between the force, deflection, and input voltage, resulting in the following:

$$F_{ext} = -k_A d + k_A d_{33} n_A u_A = -k_A d + F_u\tag{4}$$

with the piezoelectric constant d_{33} and the number of stack layers n_A of the piezoactuator. Here, the external forces including the gravitational force are summarized in F_{ext}. Due to the stiffness term, the maximum force, i.e., the blocking force F_{max}, is achieved by applying the maximum voltage U_{max} in case of a zero displacement $d = 0$. By inserting these values into (4), we obtain the following expression:

$$k_A d_{33} n_A = \frac{F_{max}}{U_{max}}\tag{5}$$

and thus simplify the piezoelectric force as the following:

$$F_u(t) = \frac{F_{max}}{U_{max}} u_A(t).\tag{6}$$

The magnet motion can be summarized to the following:

$$m\ddot{z} = \sum_{k=1}^{m} F_{pm,k}(z) + \sum_{k=1}^{n} F_{em,k}(z, i_k) + F_c(z, \dot{z}, d, \dot{d}) - c_f\dot{z} - mg,\tag{7}$$

with the force $F_{pm,k}(z)$ corresponding to the k-th ring magnet, gravitation according to the proof mass m and friction with the coefficient c_f. Note that the time dependencies are neglected for readability. The magnetic forces can further be decomposed into position-dependent shape functions $\tilde{f}_{pm,k}(z)$, $\tilde{f}_{em,k}(z)$ and proportional weighting factors as the following [22,23]:

$$F_{pm,k}(z) = B_{r,pm,k}\, B_{r,p}\, \tilde{f}_{pm,k}(z) = f_{pm,k}(z)\tag{8}$$
$$F_{em,k}(z, i) = i_k\, B_{r,p}\, \tilde{f}_{em,k}(z) = i_k\, f_{em,k}(z).\tag{9}$$

The values $B_{r,p}$, $B_{r,pm,k}$ are the magnetic remanences of the proof mass and ring magnets. The forces are precomputed numerically with ANSYS® Maxwell [24] and are implemented as spline-interpolated lookup tables.

The contact force $F_c(z, \dot{z}, d, \dot{d})$ between the movable mass and the piezoelectric actuator is approximated by a viscoelastic formulation adapted from Specker et al. [25]. Thus, by softening the contact characteristic, we mathematically allow a small penetration $s_\perp(t) = z(t) - d(t)$ of the piezo surface being counteracted by a force. The adapted approach is comparable to a spring with nonlinear contact stiffness, or activation function, $\mathcal{R}_a(s_\perp)$, but additionally takes into account energy loss at the impact, using a power function $\mathcal{R}_p(\dot{s}_\perp)$. The contact force can then be computed by the following:

$$F_c(s_\perp, \dot{s}_\perp) = \mathcal{R}_a(s_\perp)\frac{\partial \mathcal{R}_p(\dot{s}_\perp)}{\partial \dot{s}_\perp}.\tag{10}$$

Since the original formulation of Specker et al. considers possible material breach within $\mathcal{R}_\mathrm{p}(\dot{s}_\perp)$ and $\mathcal{R}_\mathrm{a}(s_\perp)$, which are not necessary in this work, the individual functions are adapted. Thus, we use the power function as follows:

$$\mathcal{R}_\mathrm{p}(\dot{s}_\perp) = \frac{\dot{s}_\perp}{2} - \frac{\log(\cosh(r_d \dot{s}_\perp))}{2r_d} \tag{11}$$

with damping parameter r_d and the following piecewise activation function:

$$\mathcal{R}_\mathrm{a}(s_\perp) = \begin{cases} 0, & \text{if } s_\perp \geq 0 \\ \frac{k_\mathrm{c}^2}{4F^*} s_\perp^2, & \text{if } \frac{-2F^*}{k_\mathrm{c}} < s_\perp < 0 \\ -k_\mathrm{c} s_\perp - F^*, & \text{otherwise}. \end{cases} \tag{12}$$

Here, k_c and F^* are design parameters corresponding to the contact stiffness and the force at the transition between the quadratic and linear parts. These values can be adjusted to achieve realistic impact behavior while being feasible for a standard numerical ODE-solver. Finally, we define the state vector $x(t)$ comprising all individual state variables, i.e., the following:

$$x(t) = \begin{bmatrix} z & \dot{z} & i_1 & i_2 & \ldots & i_n & d & \dot{d} \end{bmatrix}^\top. \tag{13}$$

It is important to mention that the presented model is not assumed to be a highly accurate representation of the later setup. Instead, it is necessary in an early stage of system design when no prototype is available. It is deployed to analyze system properties and to design component dimensions and controller architectures.

2.2. Trajectory Planning and Control

Based on the system model, we now derive a control strategy that efficiently exploits the cooperative mechanism in order to minimize the transient time and input current. To this end, we consider the kick and catch motions as separate parts. First, the piezoelectric kick is used to accelerate the proof mass. Subsequently, the controller is switched on, transferring the new initial state, i.e., the terminal state of the kick, into the desired final state.

2.2.1. Piezoelectric Kick

Accelerating the magnet by electromagnetic actuation needs large currents. Moreover, this process is slowed down due to restrictions in the maximum current. In case of the presented actuator, we therefore use a mechanical kick force to start the upwards motion. This is achieved by applying a voltage spike to the piezoelectric actuator, resulting in a large deformation of its surface and thus a fast acceleration of the proof mass. For an efficient kick, a fast voltage rise is important, while further input after the initial acceleration is unnecessary. These requirements are fulfilled by an input voltage trajectory following the course of time expressed by the following:

$$u_\mathrm{A}(t) = \frac{U_\mathrm{p}}{\hat{U}} \left(\exp\left(-\frac{t}{\tau_1}\right) - \exp\left(-\frac{t}{\tau_2}\right) \right), \tag{14}$$

which will be used with the time constants $\tau_1 = 0.1\,\mathrm{ms}$, $\tau_2 = 2.5\,\mathrm{ms}$ in the remainder of this work. Here, $U_\mathrm{p} \leq U_\mathrm{max}$ corresponds to the predefined peak voltage, and $\hat{U}$ is a normalizing factor given by the following:

$$\hat{U} = \left(\frac{\tau_2}{\tau_1}\right)^{\tau_1(\tau_1 - \tau_2)^{-1}} - \left(\frac{\tau_2}{\tau_1}\right)^{\tau_2(\tau_1 - \tau_2)^{-1}}. \tag{15}$$

Note that for the practical setup, only the time course of the initial steep edge is important, and the exact voltage may be realized differently. Moreover, instead of an analytical approach, an experimental setup is preferable for the identification of the actual proof mass acceleration.

After the time T_{kick}, the electromagnets are switched on. We therefore use the terminal position and velocity $z(T_{\text{kick}})$, $\dot{z}(T_{\text{kick}})$ as initial values for the following control approach.

2.2.2. Flatness-Based Control

Feedback control is applied to obtain smooth motions between the resting positions and to compensate for potential disturbances. The states of the proof mass and the solenoids are independent, with exception of the impact during the downwards motion. Hence, we neglect the variables $d(t)$ and $\dot{d}(t)$ for now, and thus consider only the magnetic part. We now have to define a distinct relation as follows:

$$i_k(t) = g(F_{\text{em}}(t)) \tag{16}$$

according to (9), to compute the separate currents i_k from the overall electromagnetic force. Hence, the system is flat with respect to the output variable z, which is advantageous for the control approach. Flatness means that all states and inputs can be reconstructed by means of the output variable and a finite number of its derivatives. This enables us to choose a feasible, sufficiently smooth reference motion z_{ref} and inversely compute the state and input trajectory using this output and its derivatives [26].

It is important to mention that the relation (16) is not automatically fulfilled by the system. However, we can achieve this property by utilization of an appropriate cascade structure, which will then be combined with the flatness-based controller. For this purpose, the system is split up into two parts: the linear system comprising the solenoids, and the nonlinear part describing the relation between electromagnetic force and proof mass position. These partial systems are coupled by expression (9). The overall control approach is shown in Figure 2 and is explained in detail in the following.

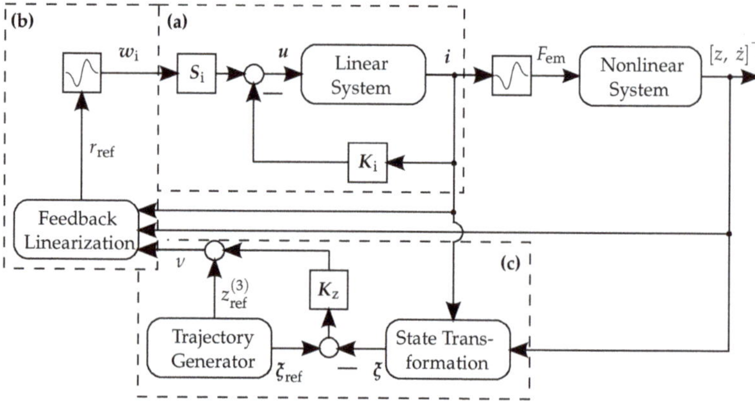

Figure 2. Combination of cascade control and flatness-based control. The linear partial system, i.e., the solenoids, are controlled by linear state feedback with controller gain matrix K_i and prefilter S_i (**a**) and the nonlinear system is handled using a feedback linearization approach (**b**). A trajectory generator is used to compute optimal reference motions and deviations between the current state and reference are compensated by a second linear controller with controller gain K_z (**c**).

In order to achieve the desired flatness property, we first have to ensure that the individual currents i_k follow a reference w_k, which fulfills the necessary condition (16). Therefore, an inner control loop for the solenoid system is used. We collect the solenoid inputs and currents within the vectors $\boldsymbol{u}$ and $\boldsymbol{i}$. According to Equations (1) and (2), the linear partial system can compactly be written as the following:

$$L\frac{d\mathbf{i}(t)}{dt} = \mathbf{u}(t) - \mathbf{R}\mathbf{i}(t),\tag{17}$$

with the invertible, symmetric inductance matrix $\mathbf{L}$ and diagonal resistance matrix $\mathbf{R}$. We can now imprint n eigenvalues $\mathbf{p}_\mathrm{i} = [p_1, p_2, \ldots, p_n]$ on the system by computing an appropriate feedback matrix $\mathbf{K}_\mathrm{i}$ by pole placement and using the control law as follows:

$$\mathbf{u}(t) = -\mathbf{K}_\mathrm{i}\,\mathbf{i}(t) + S_\mathrm{i}\mathbf{w}_\mathrm{i},\tag{18}$$

as seen in Figure 2a. Here, the subscript i denotes the correspondence to the solenoid control loop. The vector $\mathbf{w}_\mathrm{i}$ is the reference variable for the solenoid currents and S_i is a prefilter, which in our case is calculated by the following:

$$S_\mathrm{i} = \left[\mathbf{L}(\mathbf{K}_\mathrm{i} + \mathbf{R})^{-1}\mathbf{L}^{-1}\right]^{-1}.\tag{19}$$

The prefilter is used to ensure stationary accuracy. For more detailed information on multivariable linear control, the reader is referred to [27].

A second controller stabilizing the proof mass motion along a given trajectory is necessary. Due to the flatness property, dynamic feedback linearization can be used, compensating the nonlinearities within the overall system [28]. For this purpose, the following state transformation from the nonlinear state x to a linear state ξ is necessary:

$$x(t) = \begin{bmatrix} z & \dot{z} & i_1 \ldots & i_n \end{bmatrix}^\top \xrightarrow{\;\mathcal{T}(x)\;} \xi(t) = \begin{bmatrix} z & \dot{z} & \ddot{z} \end{bmatrix}^\top\tag{20}$$

This can be found by reformulating (7). So far, there does not yet exist a distinct back transformation needed for the linearization approach. In other words, given z, $\dot{z}$ and $\ddot{z}$, there exist arbitrarily many input combinations $i(t)$ leading to the same electromagnetic reference force $F_{\mathrm{em,ref}}$ and therefore to $\ddot{z}$. To overcome this, an artificial expression is derived, distinctly defining the currents $i(t)$ in dependence of a single input parameter $r(t)$, and therefore also the reversible transformation from $[z\ \dot{z}\ r]$ to $[z\ \dot{z}\ \ddot{z}]$.

Since we aim to minimize the currents, the solution at a single time instant is obtained by solving the following optimization problem:

$$\min_{i_k,\,k=1,\ldots,n} \sum_{k=1}^{n} i_k^2(t)\tag{21}$$

$$\text{s.t.} \quad \sum_{k=1}^{n} i_k(t)\,f_{\mathrm{em},k}(t) = F_{\mathrm{em,ref}}(t).\tag{22}$$

This results in the following:

$$i_k(t) = r(t)\,f_{\mathrm{em},k}(t),\tag{23}$$

with the new single input

$$r(t) = \frac{F_{\mathrm{em,ref}}(t)}{\sum_{k=1}^{n} f_{\mathrm{em},k}^2(t)}.\tag{24}$$

That means that the transformation law (23) distributes the currents in such a way that the solenoid with the largest effect on the proof mass contributes the most to achieve the necessary force. This is a large improvement, compared to the previous solution with a single electromagnet, where a singularity occurs when the movable magnet coincides with the center of the solenoid [15]. It can be seen that the desired relation (16) between the electrical currents and the electromagnetic force results from Equations (23) and (24). The transformed system then has the following dynamics:

$$\dot{\xi}(t) = f(\xi, \nu) = \begin{bmatrix} 0 & 1 & 0 \\ 0 & 0 & 1 \\ 0 & 0 & 0 \end{bmatrix} \xi(t) + \begin{bmatrix} 0 \\ 0 \\ 1 \end{bmatrix} \nu(t) \tag{25}$$

with virtual input ν corresponding to the third motion derivative $z^{(3)}$. It is now possible to find a linear control law for ν and stabilize the complete nonlinear system by the inverse computation of w_i from ν. This can be achieved by derivating (7) over time, inserting Equations (17)–(19) and subsequent reorganization. Note that for the reference currents w_i, the same transformation (23) to a single input r_{ref} has to be performed. Instead of Equation (24), this is given by the inversely computed feedback linearization as follows:

$$r_{\text{ref}}(t) = \frac{m\,\nu + c_f\,\ddot{z} - (\nabla F_{\text{pm}} + i^\top \nabla f_{\text{em}})\,\dot{z} + f_{\text{em}}^\top L^{-1}(R + K_i)\,i}{f_{\text{em}}^\top L^{-1}(R + K_i)\,f_{\text{em}}}, \tag{26}$$

which corresponds to Figure 2b. The following trajectory is now achieved by the control law:

$$\nu(t) = z_{\text{ref}}^{(3)} + K_z(\xi_{\text{ref}}(t) - \xi(t)) \tag{27}$$

with an at least twice differentiable trajectory z_{ref} that fulfills the transformed initial condition $\xi_0 = \mathcal{T}(x_0)$. For the computation of matrix K_z, we again use pole placement. The controller of the outer control loop based on the state transformation (20) is labeled as Figure 2c. Given the flatness-based controller, it now remains to find suitable trajectories $z_{\text{ref}}^{(3)}$ and ξ_{ref}, which is the goal of the following trajectory planner.

2.2.3. Optimal Trajectory Planning

We aim to find motions that efficiently enable the proof mass to switch between the equilibrium positions, while also taking into account the initial state after the kick motion. Without loss of generality, we assume an arbitrary but known initial state $\xi_0 = \xi(t_0)$ at time $t_0 = 0$. The goal is to find a sufficiently smooth trajectory, transferring the nonlinear system into the final state ξ_f over the transition time T_f.

In general, this can be achieved by connecting the initial and final state of a single transition with a polynomial, matching the initial and final conditions of the appropriate order. However, this may result in undesired oscillation and overshooting, comparable to the overfitting problem of neural networks. Instead, we use an optimization-based method with the goal to find trajectories with a low input current and short transition time, minimizing the cost function

$$J(T_f, i) = w_1\,T_f + w_2 \int_0^{T_f} i(t)^\top\,i(t)\;\mathrm{d}t \tag{28}$$

with weighting factors w_1 and w_2. To this end, we parameterize the trajectory using the parameter vector $\sigma = [\sigma_0, \sigma_1, \dots, \sigma_s]$. The optimization problem is then described by the following:

$$\min_{T_f, \sigma}\quad J(T_f, i) \tag{29}$$

$$\text{s.t.}\quad \dot{\xi}(t) = f(\xi, \nu) \tag{30}$$

$$\xi(0) = \xi_0 \tag{31}$$

$$\xi(T_f) = \xi_f, \tag{32}$$

where (30) corresponds to the linear dynamics (25), and (31) and (32) are the initial and final state conditions. The time integral in (28) is approximated by the following:

$$\int_0^{T_f} i(t)^\top i(t) \, \mathrm{d}t \approx T_s \sum_{j=0}^{q} i(j\,T_s)^\top i(j\,T_s) \tag{33}$$

with sampling time $T_s = T_f/q$. Moreover, we directly optimize the linear system input v. The minimization problem is highly nonlinear due to the relation between the electrical currents and the linear motion. It is a known issue that for nonlinear problems, gradient-based optimization algorithms strongly depend on the initial parameterization and can get stuck in unsuited local minima. Therefore, Bergman et al. [29] proposed to initialize the motion with a predefined, suboptimal trajectory. We adapt this idea by superimposing a controller-based trajectory v_{ctrl}, further denoted as nominal trajectory, with the output v_{opt} of the optimization algorithm, that is, the following:

$$v(\sigma,t) = v_{\mathrm{ctrl}}(t) + v_{\mathrm{opt}}(\sigma,t). \tag{34}$$

For the nominal trajectory, we aim to find a preferably smooth function, connecting the initial and final states. Since each individual transition, with exception of the ones including the initial kick, begins and terminates in a state with zero velocity and acceleration, we choose a 5th order polynomial as the nominal trajectory. In the case of initial velocities, however, this function can result in a large overshoot, making it unusable for the motions with a piezoelectric kick. In order to remedy this issue, we use the analytic expression as follows:

$$v_{\mathrm{ctrl}}(t) = -\lambda_1^3\, a\, e^{-\lambda_1 t} - \lambda_2^3\, b\, e^{-\lambda_2 t} - \lambda_3^3\, c\, e^{-\lambda_3 t} \tag{35}$$

for these trajectories, which can be found by the inverse Laplace transform of a stable, non-oscillating linear third-order system. Here, $\lambda_l > 0, l = 1,2,3$ correspond to predefined damping coefficients, and a, b, c are computed such that the initial conditions ξ_0 are fulfilled. Consequently, the initial values of the three time integrals of v_{opt} have to be zero. We know from Section 2.2.2 that the desired trajectory has to be at least twice differentiable. Thus, we can define v_{opt} to be a piecewise constant function as follows:

$$v_{\mathrm{opt}}(\sigma,t) = \sigma_s, \quad s\,T_{\mathrm{h}} \le t < (s+1)\,T_{\mathrm{h}}. \tag{36}$$

For a better understanding of such a superimposed trajectory, an example is visualized in Figure 3.

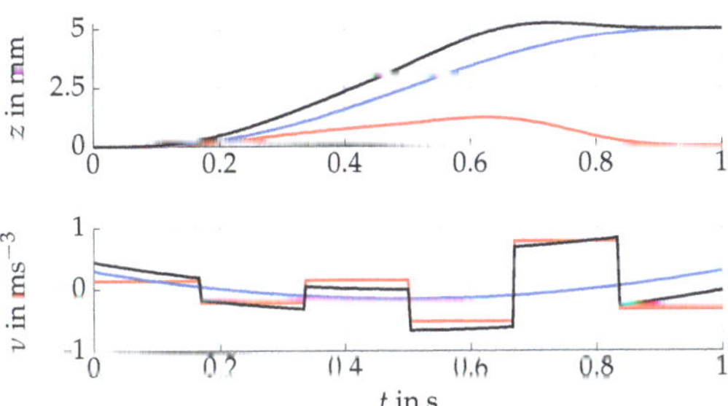

Figure 3. Superimposed proof mass trajectory that fulfills the initial and final condition up to the second order. A nominal trajectory (blue) and an optimized, piecewise constant function (red) are superimposed and result in an optimal trajectory (black). The amplitudes of the piecewise function correspond to the optimization variables σ.

The number of time intervals T_{h} can be chosen arbitrarily, but have to be at least three. This is due to the fact that the controller trajectory is used for meeting the initial conditions, but its state transition only asymptotically approaches the desired final state ξ_f. In order to compensate the deviation, the three last parameters σ_k are fixed and cannot, therefore, be optimized. Their values can be obtained by solving a system of linear equations and are therefore not discussed in detail. The advantage of the superposition-based optimization is that, in general, acceptable solutions are obtained while the parameters can simply be initialized with zeros.

However, when the proof mass comes in contact with the stack actuator, numerical issues can arise during the optimization to the lower resting position. The most acceptable results, also in terms of computational time, are achieved by computing the inverted problem, that is, optimizing the motion starting at ξ_f and ending up in ξ_0. Subsequently, the resulting state trajectory has to be mirrored with respect to time, and the sign of every even derivative must be switched before evaluating the cost function.

Finally, for the utilization of an optimization toolbox based on symbolic calculations, the lookup-table for the magnetic forces cannot be implemented. To this end, we use a single layer radial basis function (RBF) neural network to obtain an analytical approximation. See Wu et al. [30] for a more general view on RBF networks. In our case, we use the approximator as follows:

$$F_{\mathrm{pm}}(z) \approx \sum_{\mu=1}^{\eta} \alpha_\mu \, \exp\left(-\left(\frac{z - \gamma_\mu}{\beta}\right)^2\right), \tag{37}$$

which corresponds to a superposition of η Gaussian functions with respective center γ_μ and weighting factor α_μ. The parameter β adjusts the width of the basis functions and is chosen identically for each function. We then distribute the center points equidistantly throughout the relevant z-space. Since Equation (37) linearly depends on the remaining parameters α_μ, we can use linear regression to fit the approximator to the data. The same method can be used for $f_{\mathrm{em},k}$.

3. Results

We now combine the theoretical methods of the previous sections and demonstrate their usability. To this end, we first discuss the multistability and expandability property of the microactuator. This will facilitate the understanding of the subsequent simulation results of the control approach. All results were obtained using MATLAB® [31], with the exception of the magnetic forces, which were computed with ANSYS® Maxwell [24]. The optimization problem was solved using CasADi [32] with the solver IPOPT [33].

3.1. Multistability Consideration

The property of a magnet to either attract or oppose another magnet can be used to generate stable equilibrium positions, as far as the moving magnet is constrained to a single DoF motion, as is enforced by the glass tube in our case. An equilibrium position results from the overall force acting on the proof mass, being equal to zero. For a vertical setup, this corresponds to the intersection of the magnetic force with the absolute value of the gravitational force of the proof mass. Moreover, this resting position is stable if the force gradient at this point is negative, i.e., a slight deviation of the proof mass to a higher position leads to a negative force and vice versa. The stability can be proved by Lyapunov's direct method [34]. Provided that the maximum magnetic force is larger than the gravitational force, there exists at least a single stable equilibrium position. In our previous work [15], a second stable position was generated due to the contact with the stack actuator. Here, we take advantage of the nonlinearity of the magnetic force to generate a larger number of equilibria at predefined positions by superimposing magnetic fields of multiple ring magnets. For this purpose, we use a cylindrical proof mass with 1 mm height and diameter. The magnets' dimensions used in this study are given in Table 1.

Table 1. Magnet dimensions used in this study.

Component	Height in mm	Inner Radius in mm	Outer Radius in mm	Mass in mg
Proof mass	1	–	0.5	3.77
Ring magnet	0.5	1.5	2.2	–

In general, the number of generated equilibria corresponds to the number of ring magnets used in the setup. Our goal is to imprint resting positions at integral multiples of 2.5 mm for a cylindrical permanent magnet with 1 mm height and diameter. For a design comprising two ring magnets (setup 1), the resulting magnetic force is visualized in Figure 4a. The underlying setup parameters are given in Table 2.

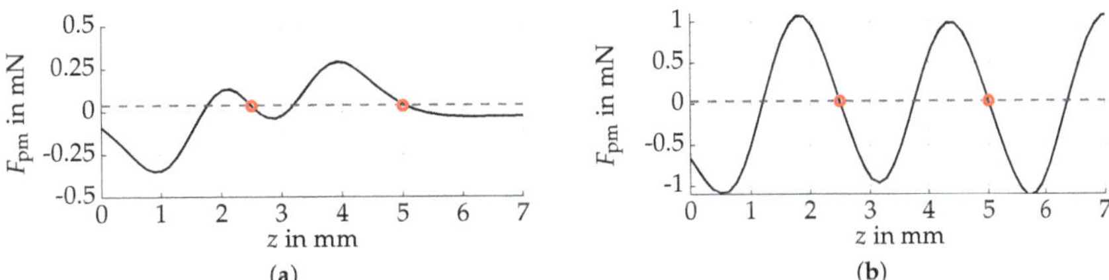

Figure 1. Simulated magnetic force characteristic F_{pm} of the superimposed magnetic fields. The intersections with the gravitational force (dashed, gray) correspond to the predefined, stable equilibria and are marked by red circles. (**a**) Setup 1: two equilibria are generated by superimposing two magnetic fields. (**b**) Setup 2: the identical resting positions are achieved by three ring magnets, but the choice of magnet parameters is more flexible. The third equilibrium of Setup 2, which is above 7 mm, is unintended and therefore not shown.

It is important to mention that the larger the ring magnets' dimensions and magnetic remanences, the larger the minimum possible distance between the two equilibrium positions. This results from the fact that, otherwise, a ring magnet generating a stable equilibrium at one position simultaneously implies a force at the desired lower position that may be either too large or too negative to be canceled by the lower magnet. This makes a systematic search for a feasible design difficult; we were able to find only a few feasible solutions for Setup 1, using unusually small remanence values (see Table 2). Similar to our previous work [15], this problem can be mitigated by utilization of an additional magnet, which is not used to generate another equilibrium, but to weaken the field of the highest magnet. Thus, we could be more flexible by the choice of the magnetic remanences. Setup 2 in Figure 4b demonstrates the advantage of this approach. Here, the two predefined resting positions were achieved by using identical ring magnets with remanences of 0.2 T, which corresponds to the lowest possible value for hard ferrite and is therefore considered producible. For Setup 1, the design is not feasible for such remanence values.

It is possible to extend the presented actuator to more equilibrium positions by stacking a larger number of ring magnets. Two examples are shown: Figure 5a shows three equilibria generated by four ring magnets, and in Figure 5b, four resting positions are imprinted by five magnets.

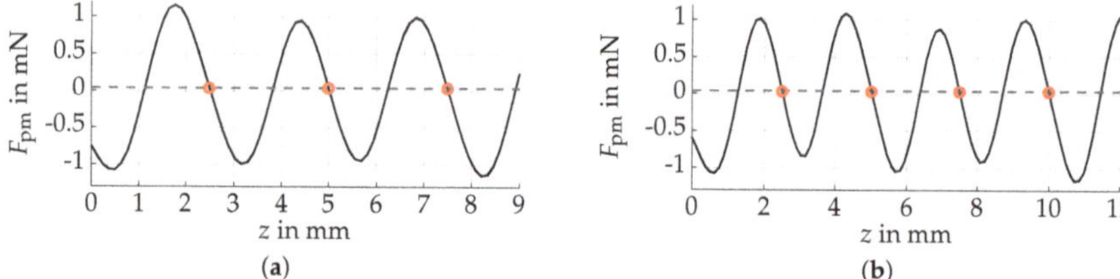

Figure 5. Magnetic force characteristic for the extended, multistable microactuator. (**a**) Setup 3: Three stable equilibria (red circles) were generated by four ring magnets. (**b**) Setup 4: Four resting positions could be achieved with five ring magnets.

We are not aware of a general limitation for the maximum number of equilibria that can be generated, but assume that it depends on the specific magnet parameters. It should be noted, however, that finding a suitable setup is not a trivial task and may not be achieved heuristically for larger setups. An optimization-based search may, therefore, be used to reduce this effort, but its solution strongly depends on a well-chosen a priori estimation of the individual magnet positions, which is generally difficult to achieve.

Moreover, the small size of the ring magnets can pose a challenge. We assume that similar results as those presented in this work can be obtained by generating weaker magnetic fields using different magnets, e.g., by replacing the small ring magnet by a larger one with internally alternating magnetizations, comparable to Halbach arrays [35], or an axis-symmetric distribution of smaller magnets. These alternative concepts are visualized in Figure 6 and should be investigated in further research. For the simulation results of the control approach in the following section, Setup 1 given in Table 2 is used.

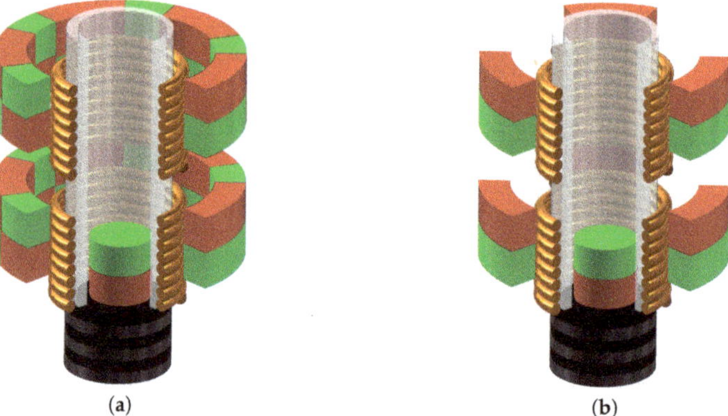

Figure 6. Alternative concepts for achieving suitable magnetic force characteristics. (**a**) A Halbach array-like magnet design may be used to weaken the magnetic field. (**b**) Multiple single magnets can be used instead of a small ring magnet.

Table 2. Experimental Setups 1–4 and respective parameters used to generate multiple equilibrium positions.

Setup	Magnet	Remanence in T	Position in mm
1	$B_{r,p}$	0.1	–
	$B_{r,pm,1}$	0.12	1.592
	$B_{r,pm,2}$	0.1	3.296
2	$B_{r,p}$	0.2	–
	$B_{r,pm,1}$	0.2	1.225
	$B_{r,pm,2}$	0.2	3.746
	$B_{r,pm,3}$	0.2	6.338
3	$B_{r,p}$	0.2	–
	$B_{r,pm,1}$	0.2	1.143
	$B_{r,pm,2}$	0.2	3.807
	$B_{r,pm,3}$	0.2	6.223
	$B_{r,pm,3}$	0.2	8.868
4	$B_{r,p}$	0.2	–
	$B_{r,pm,1}$	0.2	1.283
	$B_{r,pm,2}$	0.2	3.670
	$B_{r,pm,3}$	0.2	6.344
	$B_{r,pm,4}$	0.2	8.701
	$B_{r,pm,5}$	0.2	11.42

3.2. Simulation Study

Given the magnetic setup with two equilibrium positions (Setup 1), we now want to demonstrate the effectiveness of the cooperation between the kick and catch actuators as well as the control approach in simulation. The setup parameters are given in Table 3. The parameters concerning trajectory planning and control are summarized in Table 4, where the controller gain matrices K_i, K_z for the inner and outer control loop result from the chosen poles p_i and p_z. Note that, due to the strong dependency of the optimized trajectory from the initial velocity v_0 after the kick, it proves convenient that the parameters λ_l of the superimposed trajectory (35) also depend on the velocity. In the remainder of this work, we use the heuristically chosen law as follows:

$$\lambda = [6,\ 8,\ 10] + 2.5\,v_0, \tag{38}$$

which seems sufficient for this study. However, a more general expression may be derived by a preceding optimization, which is out of the scope of this work.

Table 3. Fixed design parameters that are used in the simulation.

Description	Value	Description	Value
Inner radius (solenoid)	0.8 mm	Resistance (solenoid)	281 Ω
Outer radius (solenoid)	1.2 mm	F_{max} (piezo)	360 N
Height (solenoid)	1.5 mm	U_{max} (piezo)	100 V
Wire diameter (solenoid)	25 μm	k_A (piezo)	1.1 MNm^{-1}
Specific resistance (solenoid)	18 nΩ m	M (piezo)	1.8 g
Number of turns (solenoid)	1222	c_A (piezo)	25
Inductance (solenoid)	3.9 mH	f (proof mass)	2.5×10^{-4}
Mutual inductance (solenoid)	2.0 μH	k_c (contact)	1.5×10^6
Position (solenoid 1)	2.2 mm	F^* (contact)	$m\,g$
Position (solenoid 2)	4.7 mm	r_d (contact)	3

Table 4. Trajectory planning and control parameters used in simulation.

Description	Value
p_i	$[-400 + 100i, -400 - 100i]$
p_z	$[-200 + 100i, -200 - 100i, -150]$
K_i	$[-280.3, \quad 1.19; \quad 0.41, \quad -279.8]$
K_z	$[7500, \quad 110, \quad 0.5] \times 10^3$
T_{kick}	$1\,\text{ms}$
β	1

Starting with the proof mass located on top of the stack actuator, we initiate the kick using different voltages between 0 V (no kick) and 100 V (full kick). Recall that the goal is to reduce the solenoid currents. Therefore, all motions are optimized with respect to the cost function (28). In the following, we use four parameters $\sigma_s, s = 1, \ldots, 4$ to describe the trajectory. The simulations are carried out for the transition of the proof mass to both the middle and upper equilibria; the results are visualized in Figure 7.

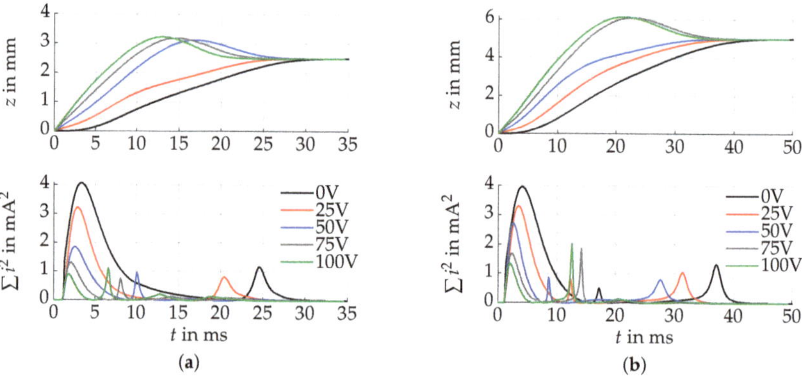

Figure 7. Combined kick and catch trajectories to both equilibria for different piezoelectric kick voltages between 0 V and 100 V. Top panel: position z of the proof mass. Bottom panel: the respective sum of squared electrical currents. (**a**) The kick to the equilibrium at 2.5 mm is shown. (**b**) A similar effect is achieved for the kick to the equilibrium at 5 mm. In both cases, the electrical currents are reduced with increasing kick impact.

It can be seen that the piezoelectric kick has a positive effect on the efficiency by largely reducing the input currents and simultaneously achieving faster transitions to both resting positions. Between the missing and full kick, the sum of squared currents is reduced by 87% for the transition to the lower resting position and by 82% for the motion to the upper equilibrium. Simultaneously, the transient time can be reduced by up to 32% and 11%, respectively. This result confirms the assumption of our previous work [15], where only a single piezo voltage and equilibrium position were used to demonstrate this effect. In the time course of the squared currents in Figure 7b, a second peak in the current can be seen for each kick. This corresponds to the current needed to decelerate the proof mass. This suggests that for larger piezo forces, the kick may have the opposite effect, due to an increased effort for damping the magnet velocity. For a general setup, the kick should therefore, be optimized simultaneously with the trajectory.

So far, we assumed that for a certain voltage, each kick is identical and corresponds to the time course predicted by our model, which in turn allows us to precisely follow the trajectory. However, the preliminary experiments showed that this is not necessarily the case, e.g., due to random impact between the proof mass and the guiding tube. This results in an unexpected initial state when the controller is switched on after the kick. In order to compensate for such disturbances, feedback control is used. Here, we approximate the effect of an initial disturbance by applying a different voltage to the piezo than what was expected within the optimization. The ability to asymptotically converge to the intended motion is demonstrated, using the 50 V-kick to the upper equilibrium. The initial state uncertainty is artificially generated by both higher and lower voltage impulses. In all simulated cases, the controller is able to compensate the disturbance within the transition time as seen in Figure 8.

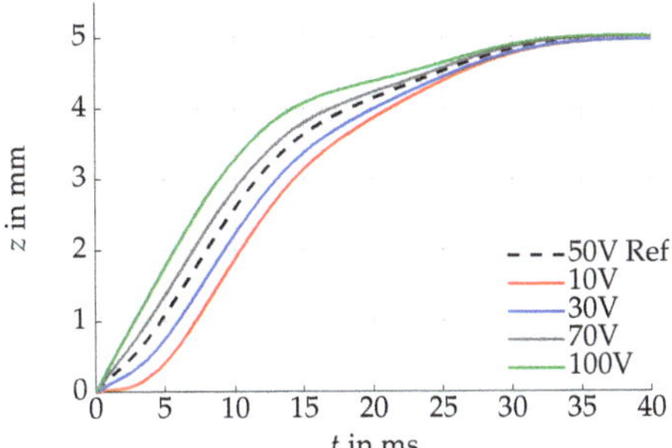

Figure 8. Feedback control simulation compensating for initial state uncertainty. The expected initial state resulting in the reference trajectory (dashed, black) for 50 V is not achieved due to incorrect kick voltages. Feedback control compensates this error and the proof mass motion asymptotically converges to the desired trajectory for different piezo voltages.

Finally, we evaluate the efficiency of the trajectory generator in terms of minimum energy effort. To this end, we compute reference motions using a different number of optimization parameters σ_s. The trajectories are then compared with the costs of the classical unoptimized approach, i.e., connecting multiple polynomials. As mentioned in Section 2.2.3, the polynomials tend toward large overshoots when combined with the piezoelectric kick. Thus, we neglect the kick and only apply electromagnetic actuation for the following comparison. We evaluate the overall costs of a motion containing all combinations of transitions between the three equilibrium positions. A visualization of the costs over the number of optimization parameters is given in Figure 9. Three of these trajectories are shown in Figure 10, namely, the unoptimized polynomial trajectory and two optimized ones with 2 and 20 optimization parameters σ_s.

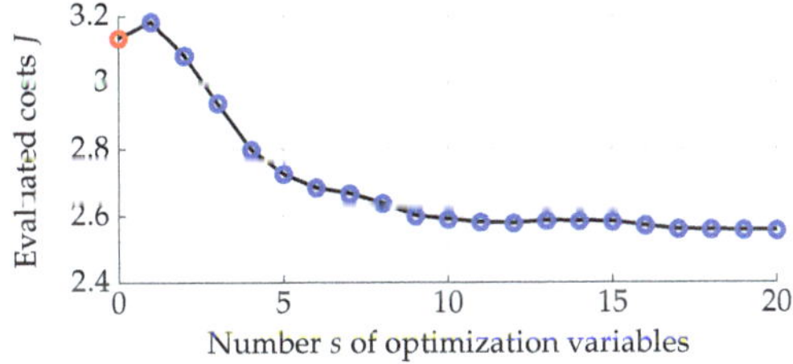

Figure 9. The trajectory costs over the number of parameters are shown. The red circle corresponds to the polynomial trajectory and the blue circles to the optimized motions. Although there is a fast cost function decrease for a low number of parameters, the costs oscillate and slightly increase for more than 10 optimization variables.

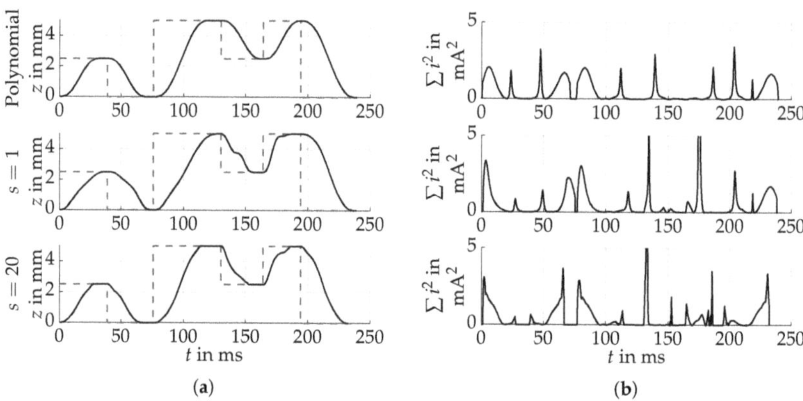

Figure 10. Comparison of trajectories and respective currents with different numbers of optimization variables. (**a**) Top panel: for the polynomial trajectory, smooth motions were generated and the transition times were optimized. Middle panel: the trajectory with two parameters σ_s and optimized transient time is less smooth but has lower costs. Bottom panel: the lowest costs were achieved with 20 parameters σ_s. The dashed line corresponds to the desired final position of the proof mass. (**b**) The sum of squared currents corresponding to the different trajectories is shown. For the trajectory with 20 parameters, the solution is non-smooth due to fast changes in the current, but the transient time and the sum of squared currents are the lowest.

Although the polynomial trajectory is the smoothest one, the optimized motions are more efficient, compared to the cost function. As was expected, the cost function values decrease with the increasing number of parameters. However, the optimized solution using a single parameter σ_s is less efficient than the time-optimized polynomial trajectory. This is due to the fact that the σ-dependent trajectory uses an exponential function for transitions from the bottom to an upper equilibrium, which is less effective than the polynomial in the case where there is no piezoelectric kick. For $s > 1$, the superposition strategy is more efficient than the polynomial trajectory. Moreover, the polynomial trajectory can result in a large overshoot when applied after the kick motion.

An interesting result corresponds to the non-smoothness of the optimized trajectories, which appears in the form of larger spikes in the electrical current. Compared to the mostly smooth input for the polynomial approach, the optimized trajectories make use of short but high current spikes in order to efficiently move the proof mass. Although the maximum input current is larger for the optimized trajectories, the evaluated time integral is smaller for the trajectory with 20 parameters, and the costs concerning the transition times are lower for both optimized trajectories. The correlation between the trajectory smoothness and the cost function should be investigated in the future. For a final assessment of the quality of the presented optimization approach, knowing the lower bound, i.e., the global minimum, is an advantage. This could be realized by a global optimizer, such as a genetic algorithm.

It should be mentioned that the nominal trajectory plays an important role for finding a meaningful solution, but may also prevent the optimization algorithm from finding a better optimum. Therefore, a considered choice should be made beforehand. Eventually, the individual motions could now be stored on the controller hardware of a real setup and run on command to initiate the respective motion. Due to its general simplicity, we assume that the presented approach can also be used as an initial solution for similar flat 1-DoF systems.

4. Discussion

The magnetic microactuator has several advantages, such as multiple, asymptotically stable equilibria and the expandability to a larger number of resting positions. This can prove useful for future applications, where large working ranges are necessary, and the proof mass is desired to remain in its position, even without a constant energy supply. We have shown that a cooperative mechanism combined with an optimal control strategy is useful for reducing the electrical load on the solenoids, which results in less heat loss and may extend the durability of the components. Moreover, the control strategy is generalizable to an arbitrarily extended design, such that the implementation effort is kept low. However, the non-smooth behavior of the optimized electrical currents may be undesirable in an experimental setup. Smoother motions can be obtained by specifying trajectories that are differentiable more often. In contrast to other stable actuator concepts, the transitions between the equilibrium positions are unstable and the controller design is indispensable. This means that for an experimental setup, sensors for measuring the solenoid currents and the proof mass position are necessary. The latter can be challenging, especially when cost-effective solutions are preferred. Therefore, principles based on induced voltage or measuring the magnetic field can be investigated.

Moreover, the proposed approach assumes fully known permanent magnetic fields without variability in their homogeinity and field strength, which is generally not the case for real magnets. While it may be possible to experimentally adapt the magnet positions to achieve multiple stable equilibria at predefined positions, such an uncertainty in the magnetic field is problematic for the proof mass motion, due to the strong model dependence of the inverse computation within the control algorithm. Although a rough estimation of the magnet tolerances for which the trajectories can be tracked with acceptable error could be achieved through extensive simulation, careful measuring of the single magnets may still be necessary before assembly. For this reason, we suggest a first estimation of appropriate magnets in simulation, experimentally identifying the corresponding real magnets, and subsequently recomputing the magnet positions and trajectories based on the obtained values. Since small changes in the magnetic field mostly affect the inverse computation of the solenoid voltages rather than the cascaded linear control loops, we assume that the choice of the controller gains can be adopted from the initial simulation without change.

Another important question concerns the possible miniaturization of such a microactuator, which is closely linked to further research on its general limitations. In this work, we showed that the magnets' dimensions and remanences are strongly coupled with the realizability of multiple equilibrium positions, particularly when short distances between them are needed. Two possible solutions are, therefore, proposed: a Halbach array-like permanent magnet or multiple smaller magnets. In both cases, the resulting force curve should be large enough to generate a robust equilibrium, but narrow in terms of little impact on the neighboring resting positions. Extensive finite element modeling will, therefore, be necessary. Finally, the efficiency of the overall system can further be improved by simultaneously optimizing both the controller and the design, such that couplings between them are exploited. It also motivates the investigation of fast, nonlinear optimal control methods, which can enrich both control theory and microactuator technology.

5. Conclusions

In this work, we presented the extension of a cooperative, magnetic microactuator by multiple, asymptotically stable equilibrium positions. This was achieved by shaping the magnetic field by several ring magnets. Additional solenoids were used to allow motions of a magnetic proof mass over a large working range. It was shown that, in general, the setup is expandable to multiple, asymptotically stable equilibria at predefined positions. For the realization of the underlying magnetic fields, two alternatives, besides small ring magnets, were proposed to be investigated in future work. Moreover, we presented a control strategy that ensures efficient switching of the proof mass between the resting positions. This approach was designed in such a way that it is generalizable to an arbitrary number of solenoids and minimizes the input current needed by the proof mass to follow a given reference trajectory. This was also supported by an optimal trajectory planning approach that considers a cooperative interaction between a piezoelectric stack actuator and the solenoids. We assume that the actuator principle and control approach are beneficial for both the microactuator and control engineering community. However, the presented model corresponds to an approximation and may be limited to hard magnetic materials with small mutual influence on the respective magnetization. Moreover, the model parameters are critical for an accurate compliance with the actual setup. It is, therefore, important to adapt the model to the measured data. For this purpose, the presented theory will be verified based on an experimental system in future work, upon which further investigations of the limitations of the approach can be carried out.

Author Contributions: Conceptualization, C.A. and T.B.; funding acquisition, C.A. and T.B.; methodology, M.O.; software, M.O.; investigation, M.O. and A.S.; writing—original draft preparation, M.O.; visualization, M.O.; supervision, C.A. and T.B. All authors have read and agreed to the published version of the manuscript.

Funding: This research was funded by the German Research Foundation (DFG) within the priority program SPP2206 "KOMMMA", project "Kick and Catch-Cooperative Microactuators for Freely Moving Platforms".

Conflicts of Interest: The authors declare no conflict of interest.

References

1. Mita, M.; Arai, M.; Tensaka, S.; Kobayashi, D.; Fujita, H. A Micromachined Impact Microactuator Driven by Electrostatic Force. *J. Microelectromech. Syst.* **2003**, *12*, 37–41. [CrossRef]
2. Min, H.J.; Lim, H.J.; Kim, S.H. A new impact actuator using linear momentum exchange of inertia mass. *J. Med. Eng. Technol.* **2002**, *26*, 265–269. [CrossRef]
3. Peng, Y.; Liu, L.; Zhang, Y.; Cao, J.; Cheng, Y.; Wang, J. A smooth impact drive mechanism actuation method for flapping wing mechanism of bio-inspired micro air vehicles. *Microsyst. Technol.* **2018**, *24*, 935–941. [CrossRef]
4. Huang, W.; Sun, M. Design, Analysis, and Experiment on a Novel Stick-Slip Piezoelectric Actuator with a Lever Mechanism. *Micromachines* **2019**, *10*, 86 . [CrossRef] [PubMed]
5. Breguet, J.M.; Pérez, R.; Bergander, A.; Schmitt, C.; Clavel, R.; Bleuler, H. Piezoactuators for Motion Control from Centimeter to Nanometer. In Proceedings of the IEEE/RSJ International Conference on Intelligent Robots and Systems, Takamatsu, Japan, 31 October–5 November 2000; pp. 492–497.
6. Hernando-García, J.; García-Caraballo, J.L.; Ruiz-Díez, V.; Sánchez-Rojas, J.L. Motion of a Legged Bidirectional Miniature Piezoelectric Robot Based on Traveling Wave Generation. *Micromachines* **2020**, *11*, 32 . [CrossRef] [PubMed]
7. Ruiz-Díez, V.; Hernando-García, J.; Sánchez-Rojas, J.L. Linear motors based on piezoelectric MEMS. *Proceedings* **2020**, *64*, 9. [CrossRef]
8. Floyd, S.; Pawashe, C.; Sitti, M. An Untethered Magnetically Actuated Micro-Robot Capable of Motion on Arbitrary Surfaces. In Proceedings of the IEEE International Conference on Robotics and Automation, Pasadena, CA, USA, 19–23 May 2008; pp. 419–424. [CrossRef]
9. Dieppedale, C.; Desloges, B.; Rostaing, H.; Delamare, J.; Cugat, O.; Meunier-Carus, J. Magnetic bistable micro-actuator with integrated permanent magnets. In Proceedings of the IEEE Sensors, Vienna, Austria, 24–27 October 2004; pp. 493–496. [CrossRef]
10. Stepanek, J.; Rostaing, H.; Lesecq, S.; Delamare, J.; Cugat, O. Position Control of a Levitating Magnetic Actuator Applications to Microsystems. In Proceedings of the 16th Triennial World Congress, Prague, Czech Rebubic, 4–8 July 2005; pp. 85–90. [CrossRef]
11. Ruffert, C.; Gehrking, R.; Ponick, B.; Gatzen, H.H. Magnetic Levitation Assisted Guide for a Linear Micro-Actuator. *IEEE Trans. Magn.* **2006**, *42*, 3785–3787. [CrossRef]
12. Ruffert, C.; Li, J.; Denkena, B.; Gatzen, H.H. Development and Evaluation of an Active Magnetic Guide for Microsystems With an Integrated Air Gap Measurement System. *IEEE Trans. Magn.* **2007**, *43*, 2716–2718. [CrossRef]
13. Laurent, G.J.; Delettre, A.; Zeggart, R.; Yahiaoui, R.; Manceau, J.F.; Le Fort-Piat, N. Micropositioning and Fasst Transport Using a Contactless Micro-Conveyor. *Micromachines* **2014**, *5*, 66–80. [CrossRef]
14. Poletkin, K.; Lu, Z.; Wallrabe, U.; Korvink, J.G.; Badilita, V. Stable dynamics of micro-machined inductive contactless suspensions. *Int. J. Mech. Sci.* **2017**, *131*, 753–766. [CrossRef]
15. Olbrich, M.; Schütz, A.; Kanjilal, K.; Bechtold, T.; Wallrabe, U.; Ament, C. Co-Design and Control of a Magnetic Microactuator for Freely Moving Platforms. *Proceedings* **2020**, *64*, 23. [CrossRef]
16. Schütz, A.; Hu, S.; Rudnyi, E.B.; Bechtold, T. Electromagnetic System-Level Model of Novel Free Flight Microactuator. In Proceedings of the 21st International Conference on Thermal, Mechanical and Multi-Physics Simulation and Experiments in Microelectronics and Microsystems, Cracow, Poland, 5–8 July 2020; pp. 1–6. [CrossRef]
17. Schütz, A.; Olbrich, M.; Hu, S.; Ament, C.; Bechtold, T. Parametric system-level models for position-control of novel electromagnetic free flight microactuator. *Microelectron. Reliab.* **2021**, *119*, 1–9. [CrossRef]

18. Goldfarb, M.; Celanovic, N. A lumped parameter electromechanical model for describing the nonlinear behavior of piezoelectric actuators. *J. Dyn. Sys. Meas. Control* **1997**, *119*, 478–485. [CrossRef]
19. Gomis-Bellmunt, O.; Ikhouane, F.; Castell-Vilanova, P.; Bergas-Jane, J. Modeling and validation of a piezoelectric actuator. *Electr. Eng.* **2007**, *89*, 629–638. [CrossRef]
20. Richter, H.; Misawa, E.A.; Lucca, D.A.; Lu, H. Modeling nonlinear behavior in a piezoelectric actuator. *Precis. Eng.* **2001**, *25*, 128–137. [CrossRef]
21. Main, J.A.; Garcia, E. Piezoelectric Stack Actuators and Control System Design: Strategies and Pitfalls. *J. Guid. Control Dynam.* **1997**, *20*, 478–485. [CrossRef]
22. Agashe, J.S.; Arnold, D.P. A study of scaling and geometry effects on the forces between cuboidal and cylindrical magnets using analytical force solutions. *J. Phys. D Appl. Phys.* **2008**, *41*, 1–9. [CrossRef]
23. Ravaud, R.; Lemarquand, G.; Babic, S.; Lemarquand, V.; Akyel, C. Cylindrical Magnets and Coils: Fields, Forces, and Inductances. *IEEE Trans. Magn.* **2010**, *46*, 3585–3590. [CrossRef]
24. Ansys Inc. *ANSYS Electronics Desktop, Release 2020 R1*; Ansys Inc.: Washington County, PA, USA, 2020.
25. Specker, T.; Buchholz, M.; Dietmayer, K. Dynamical Modeling of Constraints with Friction in Mechanical Systems. In Proceedings of the 8th Vienna International Conference on Mathematical Modelling, Vienna, Austria, 18–20 February 2015; pp. 514–519.
26. Fliess, M.; Lévine, J.; Martin, P.; Rouchon, P. Flatness and defect of non-linear systems: introductory theory and examples. *Int. J. Control* **1995**, *61*, 1327–1361. [CrossRef]
27. Mackenroth, U. Basic Properties of Multivariable Feedback Systems. In *Robust Control Systems: Theory and Case Studies*, 1st ed.; Springer: Berlin/Heidelberg, Germany, 2004.
28. Charlet, B.; Lévine, J. On dynamic feedback linearization. *Syst. Control Lett.* **1989**, *13*, 143–151. [CrossRef]
29. Bergman, K.; Ljungqvist, O.; Linder, J.; Axehill, D. An Optimization-Based Motion Planner for Autonomous Maneuvering of Marine Vessels in Complex Environments. In Proceedings of the 59th IEEE Conference on Decision and Control, Jeju Island, Korea, 14–18 December 2020; pp. 5283–5290. [CrossRef]
30. Wu, Y.; Wang, H.; Zhang, B.; Du, K.L. Using Radial Basis Function Networks for Function Approximation and Classification. *ISRN Appl. Math.* **2012**, *2012*, 1–34. [CrossRef]
31. The MathWorks Inc. *MATLAB (R2019a)*; The MathWorks Inc.: Natick, MA, USA, 2019.
32. Andersson, J. A General-Purpose Software Framework for Dynamic Optimization. Ph.D. Thesis, Department of Electrical Engineering (ESAT/SCD) and Optimization in Engineering Center, Heverlee, Belgium, October 2013.
33. Wächter, A.; Biegler, L.T. On the implementation of a primal-dual interior point filter line search algorithm for large-scale nonlinear programming. *Math. Program.* **2006**, *106*, 25–57. [CrossRef]
34. Khalil, H.K. *Nonlinear Systems*, 3rd ed.; Prentice Hall: Upper Saddle River, NJ, USA, 1996.
35. Halbach, K. Design of permanent multipole magnets with oriented rare Earth cobalt material. *Nucl. Instrum. Methods* **1980**, *169*, 1–10. [CrossRef]

Article

Model-Based Design Optimization of Soft Polymeric Domes Used as Nonlinear Biasing Systems for Dielectric Elastomer Actuators

Sipontina Croce [1],*, Julian Neu [1], Jonas Hubertus [2], Stefan Seelecke [1,3], Guenter Schultes [2] and Gianluca Rizzello [1]

[1] Department of Systems Engineering, Department of Materials Science and Engineering, Saarland University, 66123 Saarbrücken, Germany; julian.neu@imsl.uni-saarland.de (J.N.); stefan.seelecke@imsl.uni-saarland.de (S.S.); gianluca.rizzello@imsl.uni-saarland.de (G.R.)

[2] Department of Sensors and Thin Films, University of Applied Sciences of Saarland, Goebenstraße 40, 66117 Saarbrücken, Germany; jonas.hubertus@htwsaar.de (J.H.); guenter.schultes@htwsaar.de (G.S.)

[3] Center for Mechatronics and Automation Technologies (ZeMA) gGmbH, 66121 Saarbrücken, Germany

* Correspondence: sipontina.croce@imsl.uni-saarland.de

Citation: Croce, S.; Neu, J.; Hubertus, J.; Seelecke, S.; Schultes, G.; Rizzello, G. Model-Based Design Optimization of Soft Polymeric Domes Used as Nonlinear Biasing Systems for Dielectric Elastomer Actuators. *Actuators* **2021**, *10*, 209. https://doi.org/10.3390/act10090209

Academic Editor: Kui Yao

Received: 30 June 2021
Accepted: 23 August 2021
Published: 27 August 2021

Publisher's Note: MDPI stays neutral with regard to jurisdictional claims in published maps and institutional affiliations.

Abstract: Due to their unique combination of features such as large deformation, high compliance, lightweight, energy efficiency, and scalability, dielectric elastomer (DE) transducers appear as highly promising for many application fields, such as soft robotics, wearables, as well as micro electro-mechanical systems (MEMS). To generate a stroke, a membrane DE actuator (DEA) must be coupled with a mechanical biasing system. It is well known that nonlinear elements, such as negative-rate biasing springs (NBS), permit a remarkable increase in the DEA stroke in comparison to standard linear springs. Common types of NBS, however, are generally manufactured with rigid components (e.g., steel beams, permanent magnets), thus they appear as unsuitable for the development of compliant actuators for soft robots and wearables. At the same time, rigid NBSs are hard to miniaturize and integrate in DE-based MEMS devices. This work presents a novel type of soft DEA system, in which a large stroke is obtained by using a fully polymeric dome as the NBS element. More specifically, in this paper we propose a model-based design procedure for high-performance DEAs, in which the stroke is maximized by properly optimizing the geometry of the biasing dome. First, a finite element model of the biasing system is introduced, describing how the geometric parameters of the dome affect its mechanical response. After conducting experimental calibration and validation, the model is used to develop a numerical design algorithm which finds the optimal dome geometry for a given DE membrane characteristics. Based on the optimized dome design, a soft DEA prototype is finally assembled and experimentally tested.

Keywords: dielectric elastomers; dielectric elastomer actuators; bi-stable bias; polymeric dome; soft robotics; wearables; flexible actuator; finite element simulations

1. Introduction

In recent years, dielectric elastomer (DE) transducers have attracted the interest of several researchers due to their unique mix of features [1]. A DE is composed by a thin polymer membrane of acrylics or silicone material [2,3] covered in a sandwich configuration by two compliant electrodes. The application of a high voltage to the electrodes (generally within the range 1–10 kV for a polymeric membrane with thicknesses of 10–100 μm [4]) results in a change in shape, i.e., a reduction of the membrane thickness followed by a subsequent expansion in area due to the incompressibility of the elastomer. Perline et al. explained in [5] that this actuation mechanism is due to a combination of attracting forces between charges of opposite signs on the two electrodes, which lead them closer towards each other, and repulsive forces between charges of the same sign on each electrode, which cause the surface to expand. The combined effect of those forces is usually referred to as Maxwell stress, and represents the main operating principle of DE actuators (DEAs) [6].

Another important feature that makes DEAs particularly attractive is the self-sensing, i.e., the possibility of estimating the DEA deflection during electrical actuation by measuring the membrane electrical capacitance [7,8]. These unique features, combined with other advantages such as large deformations (above 100%) [1], high energy efficiency [9], low cost, and lightweight, make DE transducers attractive for a wide range of application fields, e.g., valves [10], pumps [11,12], loudspeakers [13,14], Braille displays [15], artificial muscles [16,17], and medical systems [18,19]. More recently, the use of DE technology has also been investigated in less conventional application fields such as wearables [20,21] and soft robotics [22], as well as cooperative micro electro-mechanical systems (MEMS) [23]. In these specific cases, compliance and scalability of DE material represent key features for the development of successful applications. As a particular example, soft robotics arises from the need to replace conventional rigid robots with compliant and deformable bio-inspired materials. Soft robots can be integrated more easily in social activities and unstructured environments, while ensuring at the same time a safer interaction with humans [24]. The high flexibility, compliance, and lightweight requirements of these systems can be naturally met by DE transducers, which appear then as ideal candidates for the realization of soft actuators and sensors for soft robots. Some examples of DE-based soft robots prototypes presented in recent literature include bioinspired crawling robots [25], miniature underwater vehicles [26], soft tentacle arms [27], and translucent swimming robots [28].

In all those types of systems, the actuation is made possible by the softening effect that a DE membrane undergoes upon electrical activation, which results in a stroke from the undeformed to the deformed configuration. In particular, the amount of stroke that a DEA undergoes is determined by the type of biasing system which is coupled to the membrane. A comparison between different types of bias elements has been presented by Hodgins et al. in [29,30]. In these works, it is shown that coupling the DE membrane with a negative-rate biasing spring (NBS), designed in such a way its force–displacement curve properly matches the mechanical characteristics of the DE, allowing the achievement of a significantly larger stroke compared to the one obtained with conventional biasing masses and linear springs. A mechanical system exhibiting the features of a NBS can be manufactured in several possible ways (cf. Figure 1), i.e., as a pre-compressed steel cross [31,32], a pre-compressed buckled beam [33], a pair of attracting permanent magnets [34], or via more complex types of mechanisms [35]. Even though currently available NBS elements allow remarkable improvement of the stroke performance of the overall DEA system, they are all fabricated based on rigid materials (e.g., steel) and components (e.g., clamping elements, connectors). Consequently, the resulting actuators end up losing all the flexibility and high compliance advantages of DE technology and, in turn, become unsuitable for application fields such as soft robotics and wearables. In addition, rigid biasing elements turn out to be hard to miniaturize, thus preventing the integration of DE transducers within MEMS technologies. The impossibility of using currently available types of NBS in those application fields unavoidably restricts the high potential of DE technology. To overcome this limitation, innovative NBS elements need to be developed such that, when coupled with a DE membrane, provide large strokes without affecting the compliance and flexibility of the overall actuator system. In addition to that, it would be desirable to understand how to properly optimize the design of such biasing systems (e.g., in terms of geometry) to adapt them to a given DE membrane, in order to maximize the stroke performance of the resulting actuator.

Three-dimensional polymeric domes represent a possible way to overcome the above issue. When properly designed, those domes present the characteristics of a NBS, while keeping an overall compliant and flexible structure. Therefore, they appear as highly suitable to develop soft and large-stroke DEA systems. Although the feasibility and characteristics of those types of domes have been previously proven by Madhukar et al. [36] and Alturki and Burgueño [37], their use as a biasing elements for DE membranes has been proposed only recently [38]. In particular, in [38] we performed a systematic characterization of a dome-like NBS, and used it to develop the first prototype of actuator concept which showcases the feasibility of the design. To properly optimize the stroke of such kind of DEAs, the dome must be designed so that its negative-rate (i.e., stiffness) region matches the positive stiffness of the DEA. If the relationship between the free design parameters of the dome (i.e., its geometry) and the shape of its mechanical characteristics is properly understood, it can be exploited to properly optimize a dome design for a given DE membrane, thus allowing full exploitation of the true potential of compliant DEA systems.

Pre-compressed buckled beams [33]   Attracting permanent magnet [34]   Buckling polymeric dome [38]

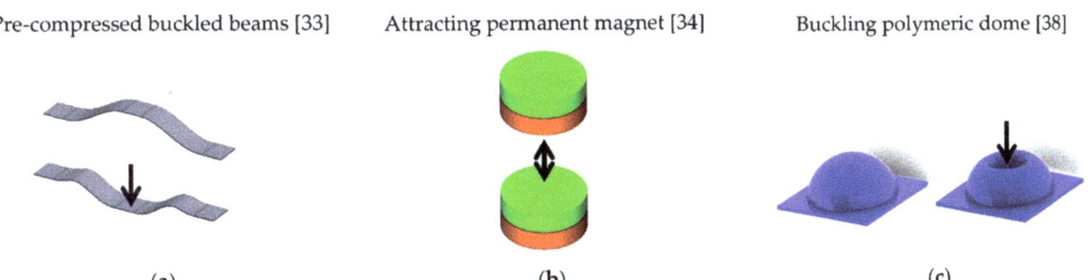

(a)   (b)   (c)

Figure 1. Examples of NBS elements: (**a**) pre-compressed buckled beam; (**b**) attracting permanent magnet; (**c**) buckling polymeric dome.

In this paper, we propose a novel approach that allows optimization of the design of the polymeric dome NBS for DEAs. The method relies on a finite element (FE) model of the dome, which allows an accurate description of the relationship between its geometry and the resulting mechanical response. FE tools are selected, as they represent a powerful means to simulate complex structures, to improve their weaknesses, and to understand their behavior in response to different external loading conditions [39,40]. After presenting the model, its experimental calibration and validation are performed, based on a number of pre-defined dome geometries. Then, the model-based design algorithm is discussed, and an example of its application is presented based on a real-life case study. After the ideal geometry of the dome is determined, the optimized DEA prototype is fabricated, assembled, and experimentally characterized. We also remark that this work represents an extended and refined version of the preliminary results presented in [41], by including a detailed discussion of the FE approach with a wider validation range, a different and more effective performance optimization approach, as well as an experimental validation based on an assembled real-life prototype. The obtained results will allow, in a future stage of our research, the development of cooperative arrays of micro-scale actuator arrays composed of layers of novel DEs and polymeric dome NBSs. Each micro-actuator will reach large stroke (due to the novel bias system) and high flexibility without losing conductivity, due to the novel sub-micron metal and carbon thin film electrodes presented by Hubertus et al. [42]. The use of such DEA arrays in wearable smart skins and loudspeakers, haptic gloves, and microconveyors will open a new generation of flexible and compliant, cooperative, and polymer-based micro-actuator devices.

The remainder of this paper is organized as follows. Section 2 presents the operating principle of DEs, and briefly discusses the NBS dome concept. In Section 3, an FE model for the novel biasing dome is presented. Experimental validation of the dome model is discussed in Section 4, while Section 5 illustrates the model-based design algorithm together with its application to develop a functioning DEA prototype. Finally, Section 6 presents concluding remarks and future research directions.

2. DEA Operating Principle

The operating principle of DEAs, as well as the role of the biasing system in determining their performance, are discussed in this section.

2.1. DEA Operating Principle

As explained above, a DE consists of a dielectric elastomer film sandwiched between two soft electrodes, forming a flexible capacitor. Its principle of operation is based on the shape change undergone due to the electric field applied through the dielectric. Specifically, when a high voltage is applied to the electrodes, a compressive Maxwell stress σ_e is generated, quantified as follows:

$$\sigma_e = -\epsilon_r\epsilon_0\left(\frac{V}{t}\right)^2,\tag{1}$$

where

- ϵ_0 is the vacuum permittivity;
- ϵ_r is the DE relative permittivity;
- V is the applied voltage;
- t is the thickness of the dielectric film.

The stress defined by (1) squeezes the dielectric film between the electrodes, resulting in area expansion due to incompressibility of the elastomer (Figure 2).

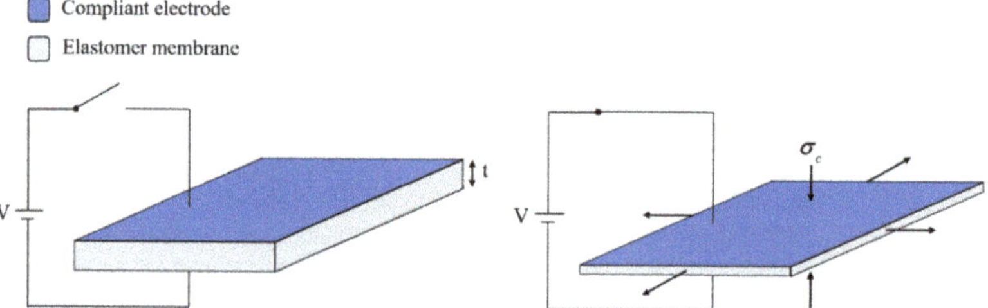

Figure 2. DE working principle: undeformed configuration on the left-hand side, deformed one due to electrical actuation on the right-hand side. Due to the combination of Maxwell stress σ_e and incompressibility of the elastomer, the reduction in thickness results in an expansion of the area.

For the manufacturing of both polymer film and electrodes, several methods and materials are available [43,44]. Silicone and acrylics represent the most typical choices for the elastomer membrane [2,3], while the electrodes are commonly based on carbon black [45]. More recently, a novel type of electrodes has been proposed by Hubertus et al. [42], consisting of thin metal sheets sputtered onto an initially pre-stretched silicone film. The main advantages of this electrode are represented by the negligible impact on the DE stiffness (due to its thickness of only 10 nm), and by the ability of maintaining a high conductivity while being stretched, thus furthering both energy efficiency and speed of the resulting actuator.

2.2. Bias Elements for DEAs

DEAs allow transformation of electrical inputs into motion by exploiting the Maxwell stress principle discussed in Section 2.1 (cf. (1)). In order to generate a stroke, a membrane DEA must be coupled with a biasing element, such as a mass or a pre-tensioned spring. When the DEA is electrically activated, the resulting Maxwell stress induces an equivalent in-plane softening within the membrane. This causes the biasing system to further pull the DE membrane until a new equilibrium state is achieved, thus resulting into a stroke. The biasing mechanism plays a crucial role for the determination of the system stroke, so its design represents a critical aspect for the performance of a DEA.

In this work we will focus on a specific DEA geometry, namely the circular out-of-plane DEA (COP-DEA) depicted in Figure 3, in both the undeformed (a) and the deformed out-of-plane (b) configurations. The actuators is based on a silicone DE membranes (Elastosil 2030 by Wacker Chemie [46]), with thickness of 50 µm and inner radius equals to 10 mm, in combination with the sputtered metallic electrodes mentioned in Section 2.1. The biasing element is connected to the passive center part of the DE membrane. When a voltage is applied, the center passive part is pushed upwards, thus generating an out-of-plane stroke.

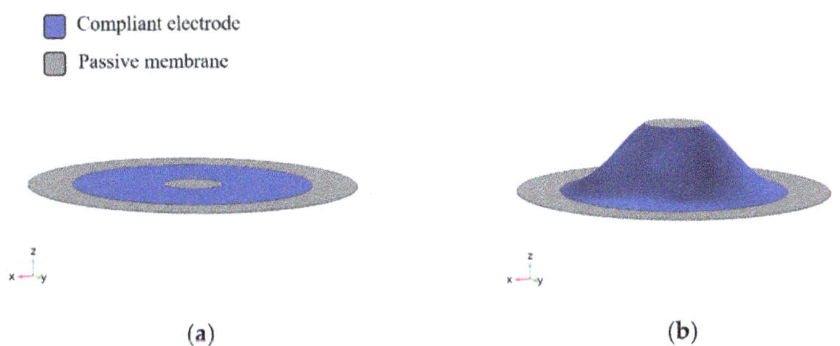

Figure 3. Three-dimensional (3D) rendering of a DEA: (**a**) undeformed flat configuration; (**b**) deformed out-of-the plane configuration.

To better clarify the importance of the biasing mechanism, as well as to understand how to quantify the stroke, Figures 4 and 5 show a cross-sectional view of the entire DEA, consisting of a COP-DEA (depicted in blue). The actuator in Figure 4 is coupled with a positive-rate biasing spring (PBS), while the one in Figure 5 is pre-loaded with an NBS (both depicted in black). To compare the strokes obtained with the two types of mechanisms, one shall consider the force–displacement characteristic curves describing the operating range of the system, sketched in the lower parts of Figures 4 and 5. In particular, the blue and red continuous lines show the behavior of the DE without and with the high voltage applied. Note how the electrical activation results in an overall softening of the elastomer membrane. The dashed curves, instead, represents the qualitative behavior of the corresponding biasing element. Note that the spring rate of the PBS appears as negative in Figure 4, while the NBS curve shows a region of quasi-linear and positive stiffness in Figure 5. This apparent change in sign of the expected slope is due to the fact that we are plotting the bias force with respect to the DE membrane displacement. Note also that, by changing the pre-stretch of the biasing element (with respect to the DE), it is possible to shift the corresponding characteristic curve along the x-axis. Notably, the gray curve represents the PBS (cf. Figure 4) and NBS (cf. Figure 5) when the DE membrane is in the undeformed configuration (cf. Figures 4a and 5a), while the black ones describe the biasing elements after that the DE membrane has been preloaded out-of-plane (cf. Figures 4b and 5)). By using a simple force equilibrium argument, it can be concluded that the intersection points between the force–displacement characteristics of the DE and the biasing mechanism determine the equilibrium states the actuator. Note that the equilibrium state explicitly depends on the applied voltage. Therefore, the horizontal distance between low and high voltage equilibrium points defines the DEA stroke. The equilibrium points reached by the whole system before and after electrical activation are called p_1, p_2 for the PBS and d_1, d_2 for the NBS, respectively. By comparing those two cases, the performance gain achieved with the NBS can readily be observed.

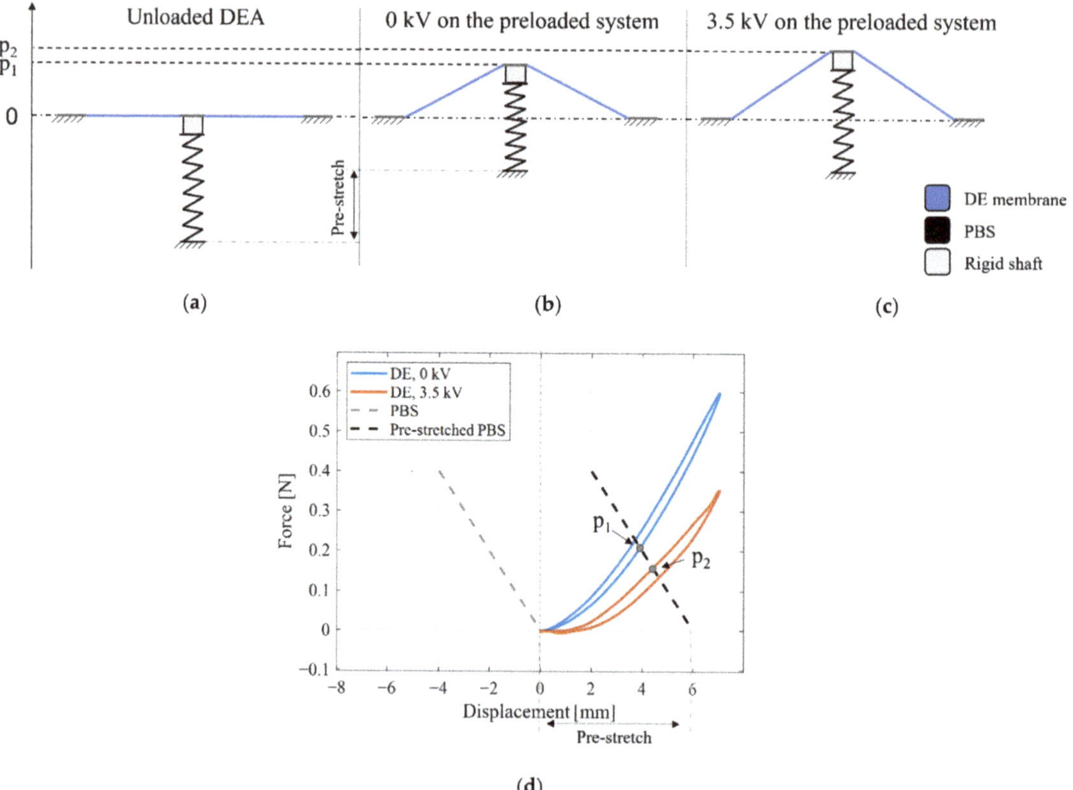

Figure 4. (**a**–**c**) Sketch of the DEA actuator composed by a DE membrane coupled with a PBS element, for three different configurations. (**d**) Actuator force equilibrium analysis, intersection points p1 and p2 represents the equilibrium states obtained by switching the voltage.

2.3. Bistable and Monostable NBS Elements

The analysis conducted in Section 2.2 is rather general and holds true for any type of biasing element whose mechanical characteristics resemble the one in Figure 5. The most common type of NBSs consist of pre-compressed metal beams, in which the negative stiffness feature is generated by means of a bi-stable behavior [33,47]. An example of this kind of system is depicted in Figure 6a. When an increasing vertical load is applied to the center point, a snap occurs which results into a switch from *State 1* to *State 2*. The corresponding force–displacement characteristic curve is depicted in the left-hand side of Figure 6a, which shows three equilibrium points in correspondence to the null external force case. The first and last ones are stable equilibria, and represent *State 1* and *State 2* configurations, while the middle one corresponds to an unstable equilibrium point. The key feature which allows use of a NBS as the biasing element for DEs is not the bi-stability itself, but rather the negative stiffness region which surrounds the unstable equilibrium point. A similar negative stiffness behavior can also be obtained by means of a monostable nonlinear structure. The monostable behavior can be achieved by as-fabricated beam designs, by combining a pre-compressed steel cross with linear spring [31], or via three-dimensional buckling polymeric domes [38]. The latter clearly represents the design solution adopted in this manuscript, due to its lack of metal parts and rigid parts. A cross-sectional sketch of this type of system, together with a qualitative example of its mechanical characteristics, is depicted in Figure 6b. Compared to the previous case, the curve intersects the horizontal axis only once, thus resulting in a single stable equilibrium point. Therefore, contrary to the previous case, the configuration *State 2* turns out to be unstable, as upon removal of the load a snap back to the initial configuration occurs.

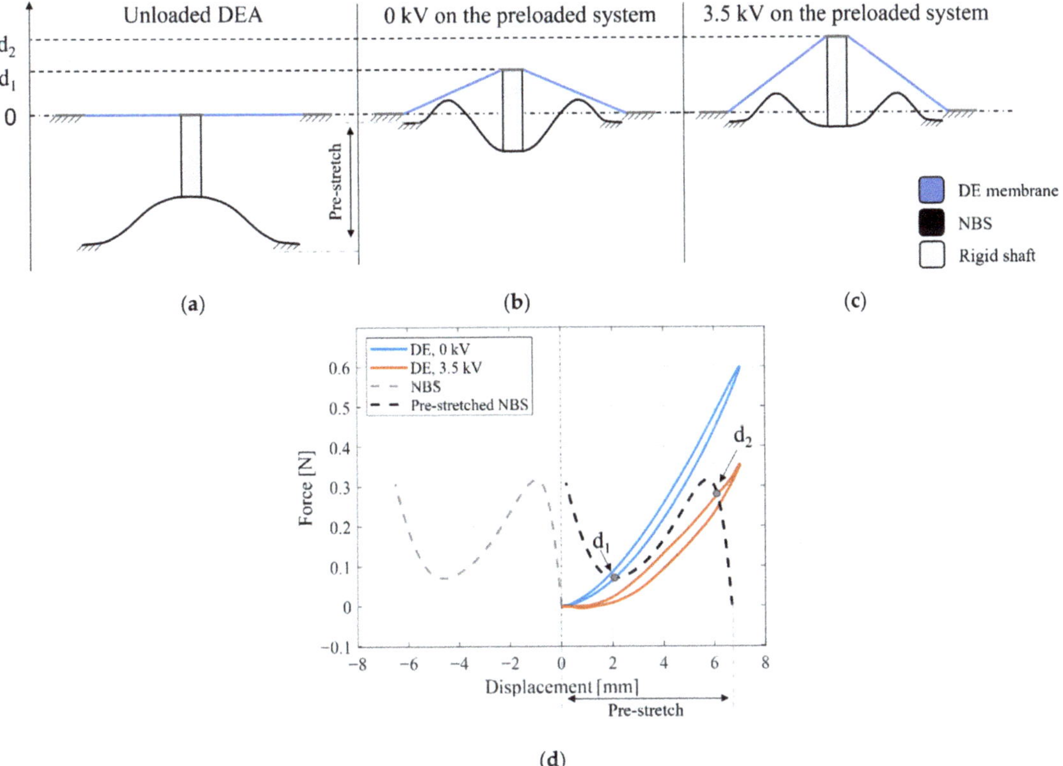

Figure 5. (**a–c**) Sketch of the DEA actuator composed by a DE membrane coupled with a NBS element, for three different configurations. (**d**) Actuator force equilibrium analysis, intersection points d_1 and d_2 represents the equilibrium states obtained by switching the voltage.

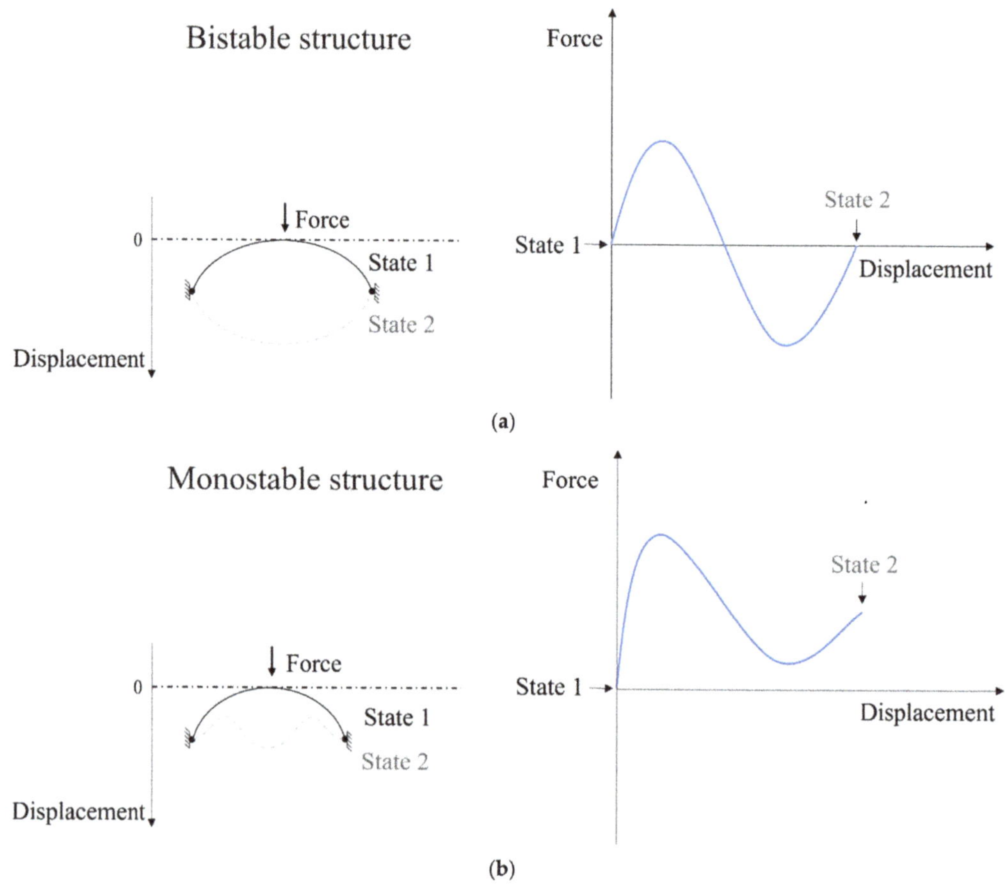

Figure 6. A sketch of (**a**) the bi-stable NBS and (**b**) the monostable structure is presented on the left-hand side, while the corresponding mechanical characteristics are shown on the right-hand side, respectively.

3. Nonlinear Biasing Dome

In this section, the FE model development for the three-dimensional fully polymeric dome is presented. The model shall predict how the dome force-deflection characteristic changes for various geometries. The FE simulation study will allow identification and improvement of the weak points of the realized structure, providing a feedback for the experimental fabrication feasibility. Moreover, the model will allow us to understand how the bias element and the DE membrane interact with each other and, in turn, to predict the performance of the coupled system for different dome geometries.

3.1. Dome FE Model

A sketch of a real-life prototype of three-dimensional polymeric dome is shown on the left-hand side of Figure 7 (for more details on the dome material and fabrication process, the reader may refer to [38] as well as to Section 4 of the present manuscript). Using the FE framework, it is possible to recreate a physics-based model of the dome experimental prototype. A 3D rendering of the FE model of the dome, implemented within COMSOL Multiphysics® environment, is shown on the right-hand side of Figure 7.

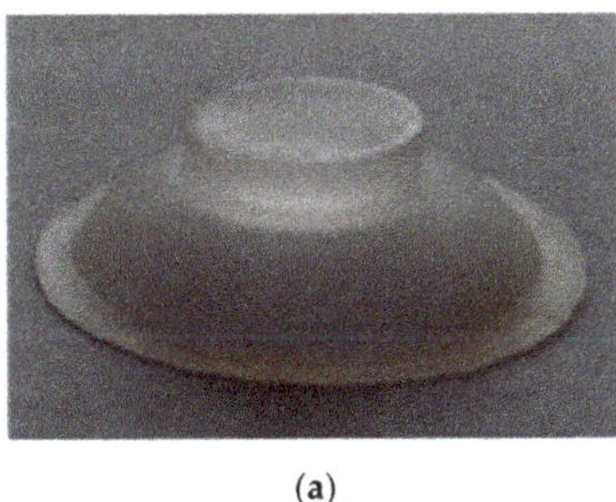
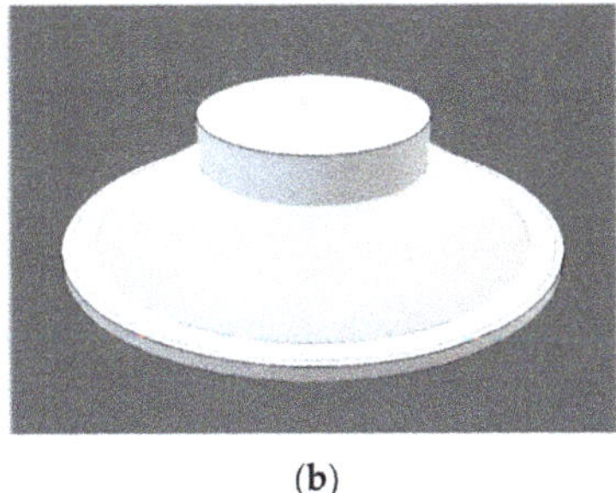

(a)   (b)

Figure 7. Fully polymeric dome: (**a**) real-life prototype; (**b**) 3D rendering implemented in COMSOL Multiphysics®.

As the dome exhibits a rotational symmetry along the vertical axis, its behavior can be equivalently described through a simple 2D axisymmetric model, shown in Figure 8. Once the 2D cross-section of the dome is implemented, a revolution around the z-axis permits generation of the complete 3D rendering, as illustrated in Figure 9. Describing the full 3D dome via an equivalent two-dimensional model allows substantial reduction of the computational complexity, without significantly affecting the significance of the result (as in our application we are solely interested in the first radial-symmetric bending mode). Based on the presented model, we will be able to perform large deformation studies by considering several geometrical configurations.

The relevant geometric parameters of the considered dome are highlighted in Figure 10. Starting from a spherical structure with wall thickness t, a flat top part with height h and radius r is added, in order to implement the connection with the DE membrane. During characterization experiments, a displacement is prescribed in the central flat part via a linear motor, and the resulting reaction force is evaluated by means of a load cell. An indenter, having also radius equal to r, is used in the experimental setup, to ensure a homogeneous distribution of the external load force on the top part (see [38] for details). An additional benefit of this design solution is the already available flat base, which offers a further support to connect the DE membrane. With regards to the FE model, the addition of this extra component allows bypassing of the simulation of the structure compressive stress through a prescribed in-point displacement, which is an important prerequisite for a correct post-buckling analysis. Indeed, the knowledge of the surface on which compression occurs facilitates the computation of the resulting force. Furthermore, the selection of a bottom base height of t_b permits avoiding taking into account tolerance errors in the dome clamping phase, resulting in more repeatable experimental results.

For concluding this section, we point out that some of the dome geometric parameters, i.e., H and r, will be left free for the optimization, while the remaining ones are fixed by design in agreement with [38]. The corresponding numerical values of the known parameters are shown in Figure 10.

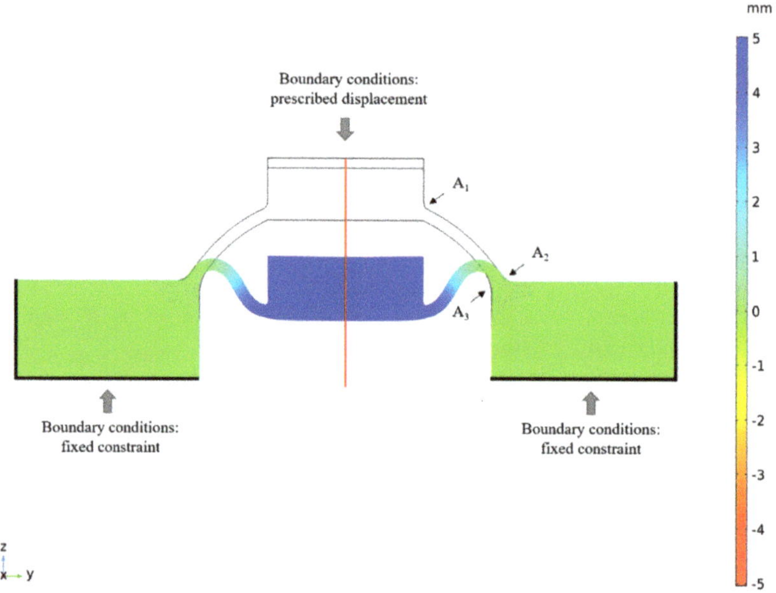

Figure 8. Two-dimensional (2D) FE model of the dome. Boundary conditions are explicitly reported as black solid lines. The non-colored image indicates the undeformed configuration, while the colored one shows the dome deformation through the colormap on the right-hand side. Note: due to radial symmetry, only the part of the dome which is located at positive y-coordinates is actually modeled.

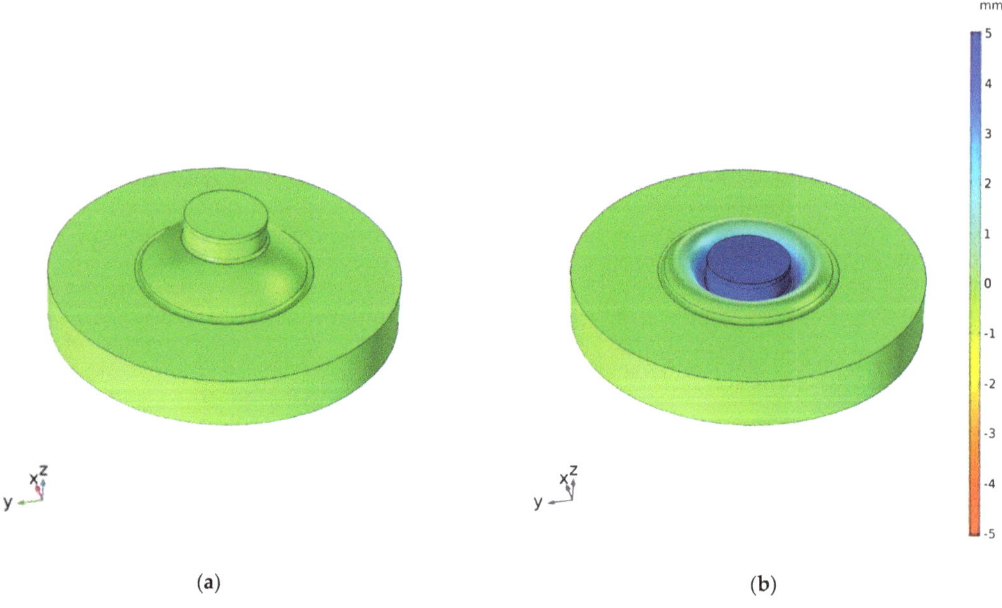

Figure 9. Three-dimensional (3D) FE of the dome, obtained from a rotation of the 2D model in Figure 8 along the *z* axis: (**a**) Undeformed configuration; (**b**) Deformed configuration.

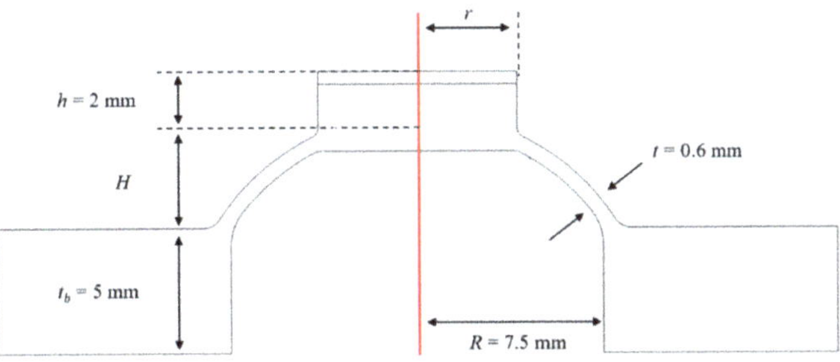

Figure 10. Two-dimensional (2D) sketch of the dome, containing the set geometric parameters.

3.2. Post-Buckling Analysis and Numerical Implementation Aspects

To properly simulate the snap-through behavior of the polymeric dome, a post-buckling analysis must be performed. In this way, it is possible to track the behavior of the polymeric dome during the applied solicitation, which will allow us to capture information about applying loads at both the critical buckling value and above. The expected shape of the force–displacement characteristic should exhibit a critical point, followed by a gradual snapping phase with a subsequent increase in stiffness. In order to perform a correct post-instability analysis, appropriate strategies must be adopted to ensure convergence of the solver. The instability represents a bifurcation problem: at the critical load, there is more than one solution. Therefore, the buckling instability will manifest numerically through an ill-conditioned or singular stiffness matrix, that results in numerical instability. On the other hand, the buckling instability represents the main feature which allows us to use polymeric domes as large-stroke biasing elements for DEs (as also discussed in Section 2). Therefore, properly account for it in the numerical simulation is of fundamental importance for DEA performance optimization.

Figure 11 illustrates an example of mechanical characteristic of the dome, based on the results in [38]. Such a curve clearly illustrates the snap-through behavior to be modeled. The critical load point (denoted as 1 in Figure 11) mathematically represents the state when the stiffness matrix becomes singular, while physically it describes the beginning of the "jump" dynamic event of the structure from state 1 to 2. Two different approaches can be adopted in FE modeling to avoid numerical instability, and thus enable solver convergence:

- Load control mode: the load force must necessarily be prescribed via nonlinear dynamic solver to solve the singularity of the problem, thus performing a time-dependent study. If a time-based analysis is conducted, there is a balance between the applied external load and elastic forces (note that all dynamic forces are neglected in this study). After that, the axial displacement represents the quantity calculated as the output. The final result is equivalent to the dashed red curve depicted in Figure 11, in which the dynamic jump from state 1 to state 2 is clearly visible.
- Displacement control mode: as the deformation represents the quantity increasing monotonically, it can be used as an input control parameter. In this way, the description of the load softening effect (occurring after the critical point) is derived unambiguously based on the simulation output.

To this end, the axial deformation is prescribed in the nonlinear static solver by setting a range of consecutive and uniformly spaced values. Consequently, the output is represented by the resulting force calculated over the same area. The final result is shown as the solid blue curve in Figure 11. The curve shows that, after the critical point is overcome, the load decreases despite the larger deformation due to a softening effect.

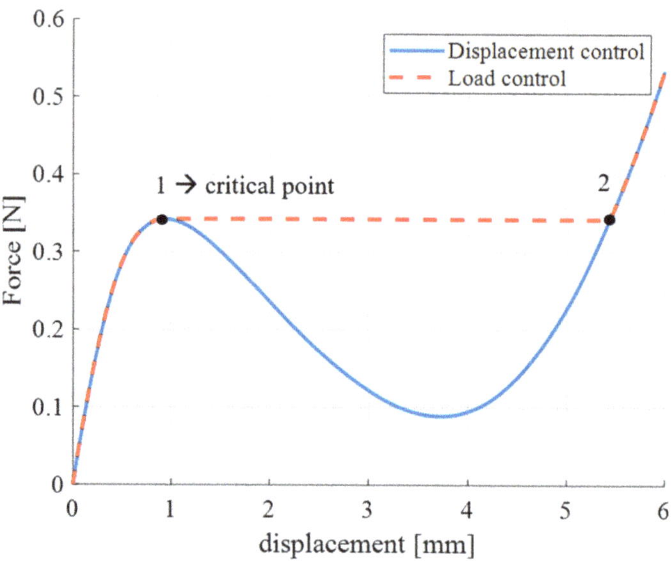

Figure 11. Experimental force–displacement curve of polymeric dome. The red and blue lines are derived by applying the load and displacement control methods, respectively. Points 1 and 2 indicate the dynamic jump occurring under load control mode, from state 1 to 2.

With the aim of fully visualizing the force values within the snap-through range, the preferred choice in our case is represented by the deformation control mode. Therefore, a prescribed displacement boundary condition is applied as input on the flat upper part, considered alongside the other boundary condition describing the fixed dome base (cf. Figure 8). Note that such boundary conditions reproduce the real constraints occurring in the experimental setup. Moreover, as we are mostly interested in the steady performance of the actuator, all the studies are conducted in quasi-static conditions to decrease the computational effort.

For simulation purpose, it is important to further define the mesh type and the size of its elements. In order to make the obtained results insensitive to the latter parameter, a force–displacement curve convergence study is performed for several planned meshes. To this end, an unstructured free triangular mesh is initially set. This type of mesh is the most widely used one for 2D models, as it represents a simple and fast way to conform at the whole geometry with localized resolutions [48]. In particular, three different types of default discretization (coarse, normal, fine) are evaluated, in order to determine the optimal one. The mesh named extremely fine is chosen as a reference solution for evaluation of the other three ones, as, due to its densely meshed design, the FE simulation converges to a tighter solution. The whole system is discretized with quadratic serendipity elements. The constitutive material behavior is modeled as described in the following. The hysteretic behavior of the material, which is observed in the experimental characterization curves, is neglected in our study in order to simplify the material description. In particular, the dome is described via a hyper-elastic material model. To account for large deformations, a nearly-incompressible Yeoh mechanical free-energy density (ψ) is selected in the solid mechanics interface [49], defined as follows:

$$\psi = \sum_{i=1}^{3} c_{i0} \left(\lambda_1^2 + \lambda_2^2 + \lambda_3^2 - 3 \right)^i + \frac{\kappa}{2}(J-1)^2. \tag{2}$$

The meaning of all the quantities in (2) is described in the following:

- Principal stretches $\lambda_1, \lambda_2, \lambda_3$;
- Material constitutive parameters c_{i0}, $i = 1, 2, 3$;
- Bulk modulus κ, which allows practical accounting of the material incompressibility in a numerically efficient way;
- Volume ratio J, equal to the determinant of the deformation gradient.

A problem encountered when simulating the considered dome geometry is that stress convergence does not occur nearby sharp corners, due to a local stress singularity in the model. This issue is numerically handled by including rounding parameters for the sharp corners, denoted as A_1, A_2, A_3 in Figure 8. The Comsol integrated direct solver MUMPS (MUltifrontal Massively Parallel Sparse) is selected, with a nonlinear Automatic (Newton) method, to solve the system of equations. The solver parameters are set as follows: maximum number of iterations of 25, initial damping factor of 10-4, and a tolerance factor of 10-2. Representative values of $c_{10} = 0.05$ MPa, $c_{20} = 1.52$ kPa, $c_{30} = 5.11$ kPa, $\kappa = 22 \times 10^6$, are chosen for the study.

The average relative percent errors, computed by comparing the reference characteristic curves with the ones obtained with different discretization methods, are depicted in Figure 12a. It is interesting to note that, even with a coarse mesh, the error is always less than 1%. This result suggests that, in future steps in which dynamic studies will be conducted on the full actuator system (i.e., biasing dome coupled with the DE membrane), the choice of a coarser mesh might lead to a good trade-off between accuracy and simulation time. On the other hand, as the extremely fine mesh requires a higher computation time for a steady-state study, we decided to set the default mesh type to normal and to customize it in the areas where the stress is more concentrated (i.e., the blue lines represented in Figure 12b). To this end, an additional element-local size is defined only along the concerned edges, ensuring a higher node density. This solution permits achievement of the same accuracy as the extremely fine mesh defined before, while also requiring less computation time. To implement this feature, we start by setting the maximum element size of the mesh equivalent to the default normal mesh value (about 1.2 mm). This parameter is then gradually decreased, until a value of 0.1 mm is reached. The last size value generates the finest possible mesh, which will be considered as reference. Afterwards, the average percent deviations between the different force–displacement curves and the reference one is quantified and shown in Figure 12c. From this study, it can be seen that the resulting curves tightly match the reference one, thus leading to a very low average percentage error for the different mesh designs. Note that, however, the smaller the maximum size of the mesh elements, the finer the resulting mesh, thus resulting in improved convergence of the relative percent error. Nevertheless, we choose a parameter-dimension value equal to 0.15 mm, as it allow mitigation of the computational time (about 1 min) while maintaining a satisfactory accuracy (percentage error smaller than 0.02%).

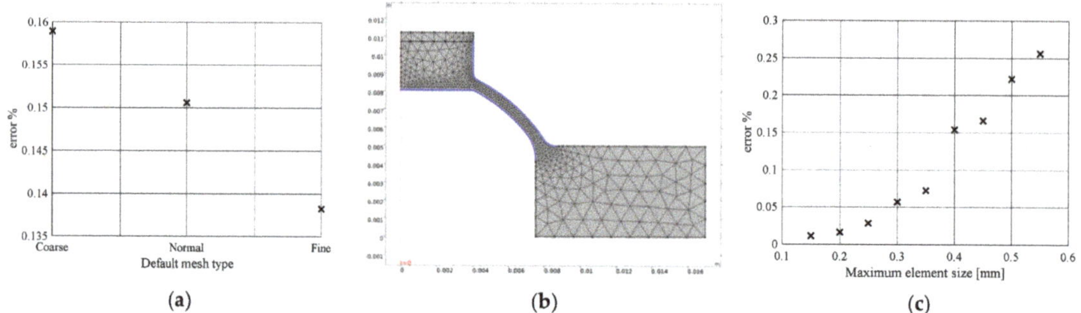

Figure 12. Mesh convergence study: (**a**) quantifies the average percentage error between the curve obtained by setting the mesh extremely fine (considered as reference) and the other default types listed on the *x*-axis; (**b**) 2D FEM, with mesh defined as normal and customized along the blue edges. (**c**) quantifies the average percentage error between the characteristics curves resulting from varying the maximum size of the mesh elements along the blue edges.

4. Dome Calibration Based on the Experimental Characterization Process

Experimental validation of the dome model presented in Section 3 is presented in this section.

4.1. Dome Experimental Characterization Process

Prior to parameter identification, an experimental characterization process is carried out on the dome, in order to obtain a dataset of repeatable force–displacement measurements. The dome is made with a WACKER SILGel® 612 EH silicone [50], through a process of casting into 3D-printed molds. The experiments are conducted by deforming the flat upper part of the dome with a 0.1 Hz sinusoidal displacement, via an indenter connected to a linear actuator (Aerotech, Inc., Model: ANT-25LA). During the deformation, the force is acquired via a load cell (ME-Meßsysteme GmbH, KD40s). In this way the force–displacement characterization curve of the dome is obtained. Due to the tolerances of the 3D printer for mold manufacture and the manual mixture of the silicone used for the domes, several batches representative of the same geometry but made with different molds are created. In this way, it is possible to quantify the reliability and repeatability of the experimental curves, which will be included in the dataset shown in the next section. For further details regarding the experimental characterization of the dome, please refer to [38].

4.2. Dome Identification and Validation

The goal of the experimental identification consists of finding the material parameters which allow reproduction of a selected subset of the experimental results. The unknown parameters are essentially the Yeoh constitutive coefficient c_{i0}, the bulk modulus κ, as well as the rounding parameters A_1, A_2, A_3 introduced for numerical robustness purpose (see Section 3.2 for details). Clearly, the numerical values of those coefficients are chosen identically for each dome geometry. Once the constitutive parameters are known, we can use the model to reproduce the experimental trends obtained for different geometric configurations of the dome structure, in order to make reliable and accurate predictions about its behavior.

Calibration of all parameters is performed based on a repeatable set of experimental data, characterized through a procedure described in the previous section. The considered dataset contains force–displacement characteristic curves for different dome geometric configurations, built through different combinations of H and r values (cf. Figure 10). The measured force–displacement curves exhibit a moderate hysteresis, which is due to the polymeric material viscoelastic behavior. As this paper represents the first attempt at model-based dome optimization for DEA systems, we decided to neglect hysteresis modeling due to its high numerical complexity (as stated in Section 3). Such a hysteresis introduces a continuous range of possible intersections between the DE and bias curves (recall the argument in Section 2.2). Among all those values, the system naturally choses the equilibrium position which minimizes the motion range and, in turn, the actuation stroke. Therefore, we expected that neglecting the hysteresis will produce an overestimation of the true performance. Keeping this aspect in mind, the original dataset is post-processed by replacing the experimental characteristic curves with interpolating polynomial functions. The corresponding experimental curves are shown as solid black lines in Figure 13. Table 1 provides an overview of the considered geometric values, and how the dataset is split for the calibration and validation steps. The different values of H are chosen according to a preliminary investigation carried out by Madhukar et al. [36], thus resulting in a final height of the dome that ensures a monostable behavior. The range for r, instead, is chosen in such a way to match the inner radius of the experimental DE [38]. Note that 3 experiments are used for calibration and 6 for validation, respectively.

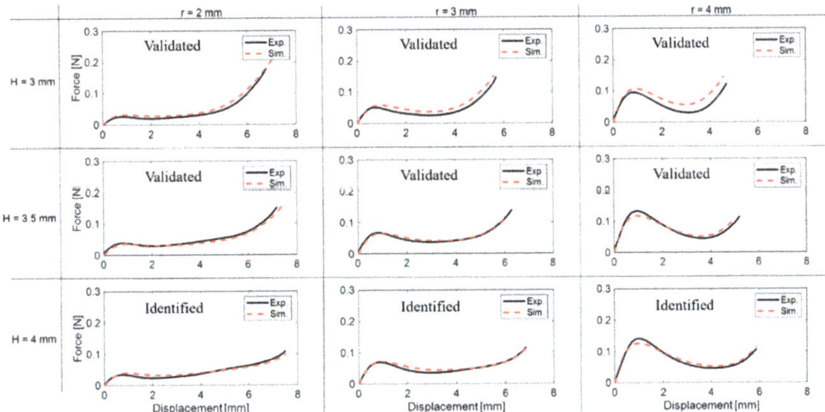

Figure 13. Matrix of plots showing the comparison among experimental (depicted in black) and simulated (depicted in red) curves, for different dome geometries.

Table 1. Experimental dataset splitting.

	r = 2 mm	r = 3 mm	r = 4 mm
H = 3 mm	Validated	Validated	Validated
H = 3.5 mm	Validated	Validated	Validated
H = 4 mm	Identified	Identified	Identified

Subsequently, the optimization module integrated in COMSOL is used for identifying the structure parameter. The Nelder–Mead optimization method is set internally, i.e., a gradient-free algorithm able to remesh the whole structure as the geometric parameters change, allowing automatic testing of different material parameter values for different model configurations [48]. The output of this step consists of those parameter values that minimize the overall variance between the experimental and simulated curves. The resulting values of the identified optimal parameters are reported in Table 2.

Table 2. Identified parameters values.

c_{10}	c_{20}	c_{30}	κ	A_1	A_2	A_3
0.11 MPa	3.29 kPa	5.73 kPa	0.25×10^6	0.44 mm	0.78 mm	2.2 mm

Afterwards, simulations of the entire dataset are performed based on the calibrated material parameters, and the results are shown as dashed black lines Figure 13. Note that a wide variability is observed among the considered experimental curves, especially concerning the shape of the negative stiffness region. Therefore, we conclude that changing both H and r represents a feasible and effective way to arbitrarily shape the characteristic curve of the dome. This fact will be of fundamental importance when optimizing the dome geometry for a given DE, as we will see in the subsequent section. Figure 13 further shows that the model allows reliable prediction of the effects of changes in geometry in each case, as the validation process describes the trend of the curves with remarkable accuracy. In particular, an average error between simulations and experiments of less than 5% is observed. We recall that our model is built upon an axial symmetry assumption. Despite that manufacturing tolerances may induce axial asymmetries in the experimental dome, the high accuracy of our predictions seems to confirm that such effects are indeed negligible. Small deviations (observed, e.g., for the $H = 3$ mm, $r = 4$ mm case) are reasonably due to manufacturing tolerances and misalignments. As a result, we conclude that our model represents an effective tool to foresee how the dome geometry affects the shape of the resulting mechanical characteristic curve.

5. Dome Design Optimization and Experimental Validation

This section provides a description of the proposed optimization procedure for the dome geometry. In this way, the stroke-optimal design configuration can be computed without the need of manufacturing several prototypes in a trial-and-error fashion. As discussed in Section 2, the intersection points between the dome force–displacement curve and the DE membrane characteristics determine the stroke of the system. To further illustrate this aspect, we consider the curves shown in Figure 14. The DE curves correspond to the ones of the experimental prototype, while a fictious black line is used to represent the bias. When switching the DE voltage from 0 kV to 3.5 kV, the DE equilibrium position switches between points A and B (cf. Figure 14). Note that the particular shape of the bias in Figure 14 is conveniently chosen in such a way to lead to a large actuation stroke. Therefore, it represents a target behavior for an optimal biasing element. In case of the considered biasing domes, this target biasing line shall be interpreted as an approximation of the overall force–displacement characteristics, more specifically of the unstable branch. The goal of the model-based design consists thus in matching the unstable portion of the dome curve as close as possible to the ideal linear behavior.

In general, an ideal biasing curve can be specified through the location of intersection points A and B, such that the line connecting them results in a large actuation stroke and, at the same time, does not intersect the DE curves multiple times. This target behavior can be described via four independent pieces of information, i.e., the coordinates of points $A = (A_x, A_y)$ and $B = (B_x, B_y)$. These variables can be uniquely related to the following performance measures, which allow equivalently describe the target biasing curve in Figure 14 in a more intuitive way:

- Stroke, computed as the distance between states A and B along the x-axis, i.e., $B_x - A_x$;
- Slope, defined as the angular coefficient of the line connecting equilibrium points A and B, i.e., $(B_y - A_y)/(B_x - A_x)$;
- Maximum force, defined as B_y;
- Horizontal shift, defined as a constant offset applied to both A_x and B_x.

Out of these four quantities, the horizontal shift can be easily modified by changing the relative distance between DE and dome, e.g., by connecting them with spacers of different lengths. The remaining three parameters, on the other hand, must be determined by properly designing the dome characteristics through geometry optimization. As a final remark, we recall that the real-life domes are characterized by a small, yet nonzero hysteresis, which is not included in our model. Therefore, for design tolerance purpose, it is convenient not to let the target biasing curve match the DE curves too tightly.

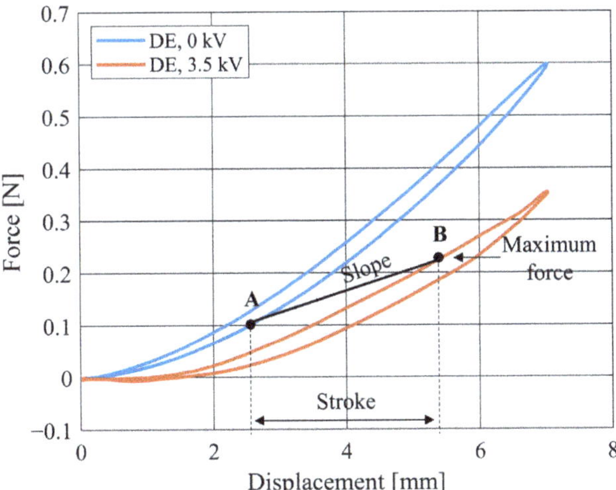

Figure 14. Intersection between the DE mechanical characteristics without (solid blue line) and with (solid red line) applied voltage, and ideal biasing behavior ensuring large stroke (solid black line) Points A e B represent the system equilibrium points for low and high voltage, respectively. Their distance along the horizontal axis defines the corresponding stroke.

5.1. Optimal Parameter Selection

The aim of this section is to find the optimal dome geometry which, starting from desired stroke, slope, and maximum force values, finds the optimal dome geometry which allows as close as possible approximation of the desired behavior. H, r, and t represent the geometric parameters that mostly affect the dome force–deformation curve (cf. Figure 10), thus they will be considered for the following optimization study.

As a first step, we formally define three functions which, given as input the dome geometry, return the desired performance parameters:

$$\begin{cases} \text{stroke} = f_1(H, r, t) \\ \text{slope} = f_2(H, r, t) \\ \text{maximum force} = f_3(H, r, t) \end{cases} \tag{3}$$

In principle, to solve (3) one should be able to freely modify all three geometric parameters. While changes in H and r can be predicted accurately by means of our FE model, the reliability for different values of t has not been investigated. This is mainly due to the practical difficulty of designing domes with different values of t, based on our manufacturing process. Therefore, we decide to keep it fixed at $t = 0.6$ mm (the same value used during identification) and optimize for H and r only. By considering (3) with $t = 0.6$ mm, and manipulating the equations under the assumption that invertibility conditions formally hold, we obtain the following system:

$$\begin{cases} H = g_1(\text{stroke}, \text{slope}) \\ r = g_2(\text{stroke}, \text{slope}) \\ \text{maximum force} = g_3(\text{stroke}, \text{slope}) \end{cases} \tag{4}$$

Thus, the values of H and r can be obtained based on the target values of stroke and slope only, while the corresponding value of the maximum force is implied by the design. The major drawback of the proposed approach is that we have no complete freedom in freely design stroke, slope, and maximum force, but we must somehow find an approximated solution for the design problem. On the other hand, working with two free parameters allows us to exploit graphical intuition to accomplish the design, as well as to simplify the numerical computation. As we will see in the following, the adopted compromise does not represent a practical issue for the design of our optimal dome geometry.

The steps needed to develop the design optimization algorithm are now summarized in the following:

1. The calibrated FE model (described in the Section 4) is used to realize a dataset of simulated force-strain curves for different dome geometries. For the considered case study, the ranges of H and r are chosen in a physically meaningful way as follows: $H \in [3, 5]$, $r \in [2, 4]$;

2. The entire design algorithm is implemented in MATLAB®, based on the obtained simulation dataset. For each simulated force-strain curve, the minimum and maximum force points defining the unstable branch of the dome characteristic are calculated and collected, in order to determine corresponding slope, stroke, and maximum force. Those minimum and maximum points are therefore considered as representatives of A and B, where the intersection with the DE characteristic curves occurs;

3. Surface fitting functions are generated to express H, r, and maximum force as a function of the stroke and slope, based on the computations performed in the previous step. Resulting functions $H = g_1(stoke, slope)$, $r = g_2(stoke, slope)$, and *maximum force* $= g_3(stoke, slope)$ are shown in Figure 15. As it can be seen, such surfaces allow unique determination of H, r, and the maximum force, once the target stroke and slope are known.

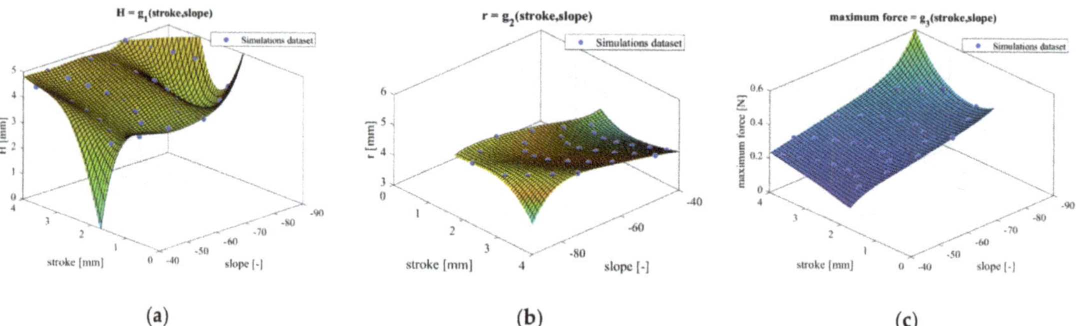

Figure 15. Surface fitting functions created based on the simulations dataset, with the final goal of deriving the dome geometric values for H and r based on the desired performance: (**a**) H as a function of stroke and slope; (**b**) r as a function of stroke and slope; (**c**) maximum force as a function of stroke and slope.

The choice to create surface fitting functions, rather than multidimensional lookup tables, is motivated by the need of increasing the computational efficiency, in conjunction with the low impact on memory storage of the former over the latter.

5.2. Design Optimization Algorithm

In this section, the steps required to develop an optimized DEA based on the above model-based procedure are reported:

1. Select a target DE membrane, and characterize it experimentally under quasi-static conditions in order to obtain the characteristic curves for minimum and maximum applied voltage;

2. Based on the obtained DE curves, estimate an ideal biasing behavior, and compute the coordinates of the corresponding intersection points A and B (cf. Figure 14);

3. Based on the coordinates of A and B, determine the desired stroke, the slope, and the maximum force values, i.e., the features that must be satisfied by the dome force–displacement curve in order to ensure the desired performance;

4. If the maximum force is not satisfactory, one can eventually start again rom point 2 and try different combinations of stroke and slope values, until an overall desirable behavior is obtained.

The output of the above procedure consists of the optimal values for H and r, which allows reproduction of a dome curve that reflects the target behavior.

5.3. Design Procedure Validation

For a better understanding of the working principle, and to verify the reliability of the design optimization algorithm, a design case study is presented in this section. The considered DE membrane is the same as the one described in Section 2. Starting from the measured COP-DEA characteristic curves (depicted in blue and red for low and high voltage, respectively), we define the desired intersection points (marked with black circles), which lead to a desired stroke of 3.6 mm. These points are then used to generate inputs for the design algorithm, which leads optimal geometry values equal to $H = 4.6$ mm and $r = 4.8$ mm. Note that those value fall outside the calibration and validation dataset shown in Figure 13, thus the optimal design relies on an extrapolation performed by our model.

Based on the obtained geometric parameters, the optimal dome characteristic curve is estimated via an FE simulation, which is depicted as a dashed black curve in Figure 16. It can readily be observed that the design goal is successfully achieved, as the simulated dome curve intersects those of the DE in the desired points initially defined. This is true despite only two free parameters have been optimized (instead of three), thus confirming the validity of the proposed approach. Based on the curves in Figure 16, a theoretical stroke of 3.8 mm can be predicted. To quantify the reliability of the design, a dome with the optimal geometry is experimentally assembled and characterized. The result is shown as a solid gray curve in Figure 16. Comparing the experimental force-deflection curves with the simulated one, we conclude that the measured curve tightly matches the simulated, thus confirming the accuracy of our model.

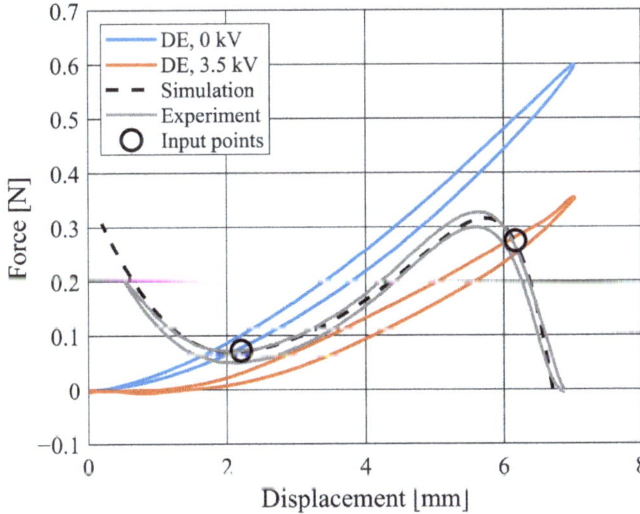

Figure 16. Design algorithm experimental validation. The input is represented by the desired intersection points (black circles), while the dashed black curve and solid gray curve represent the simulated and experimental results based on the optimized dome geometry.

The optimal dome is then coupled with the DE membrane and used to manufacture a complete actuator system. A picture of the assembled DEA device is shown in Figure 17. Manufacturing and assembly are performed similarly to [38]. In particular, before assembly, the electrical connections to the metal electrodes are achieved through a copper tape. Then, a double-sided tape cut with a cutting machine is used to couple the optimal polymer dome with the DE membrane. The assembly is performed through the use of 3D-printed tools, thus avoiding manual positioning and allowing improvement in the accuracy. As the DE is controlled with high voltage, it has been placed underneath the insulating dome in order to operate in total safety.

Figure 17. Picture of the optimized DEA prototype.

After completing the actuator manufacturing, an electro-mechanical characterization is performed with the aim of evaluating the stroke achieved for different input voltage signals. The motion of the DEA flat top is measured through a Laser (Keyence, model: LK-G87), while the analog voltage output of the voltage amplifier is monitored through NI LabVIEW module (PXI-7852). A first set of tests is conducted by considering a constant voltage step, which is applied for a total duration of 10 s. Several steps are considered, each one ranging from 0 kV up to a different maximum value ranging from 3 kV to 3.5 kV, with steps of 0.1 kV. The minimum voltage is set to 0 kV, as it corresponds to zero actuation force, while the maximum voltage value is limited by the dielectric breakdown field strength (i.e., E_{DE} = [80–100] V/μm) [46]. The results are shown in Figure 18. For the maximum voltage, a stroke of about 3.1 mm is observed. This quantity is smaller than the theoretical value of 3.8 mm predicted with the graphical method, possibly due to unavoidable tolerances, misalignments, and inaccuracies occurring during the manufacturing process.

Nevertheless, the obtained stroke is still remarkably high, thus confirming the effectiveness of our design approach. Note that, a different amount of stroke is obtained by changing the voltage amplitude, thus confirming the possibility of using the developed DEA as a proportional actuator. Further experiments are conducted by considering sinusoidal voltage waveforms having frequency of 0.1 Hz and different amplitudes, ranging from 3 kV to 3.5 kV with steps of 0.1 kV. The resulting voltage and displacement over time are shown in Figure 19, while the corresponding displacement over voltage curves are reported in Figure 20. A maximum stroke of about 3.1 mm is observed also in this case. Additionally, note that the displacement-voltage curves exhibit a hysteretic behavior, which is mainly caused by the bi-stability of the dome NBS, rather than from polymer viscoelasticity. Such hysteresis is commonly observed in NBS-DEA systems and is a direct consequence of the same mechanism which leads to the large stroke performance.

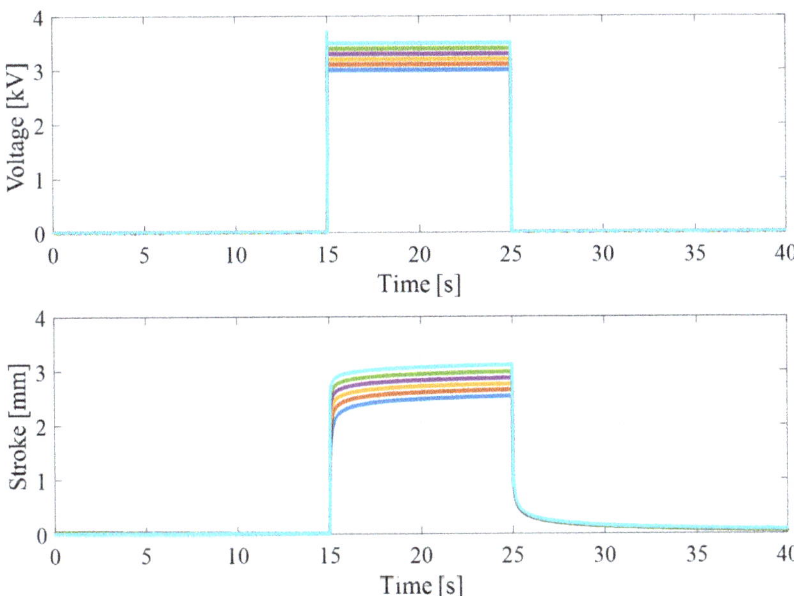

Figure 18. The upper part shows a step voltage signal applied as input, applied for a time of 10 s. The different colors represent signals of different amplitudes set as the input, which vary in the range 3–3.5 kV with increments of 0.1 kV. In the lower part, the corresponding stroke reached by the DEA is depicted.

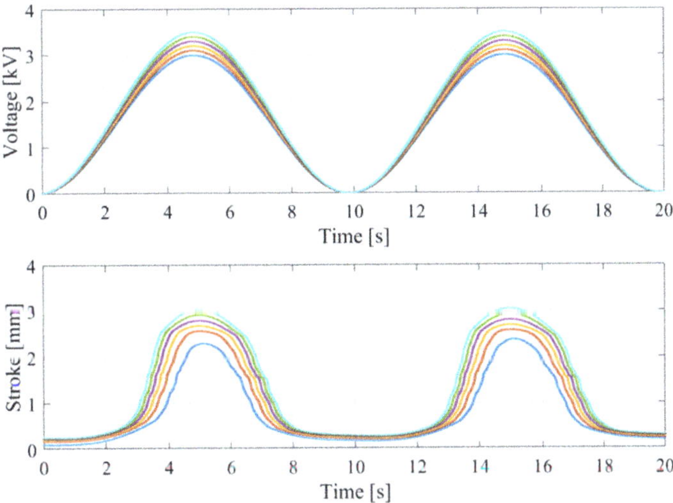

Figure 19. The upper part shows a sinusoidal voltage signal applied as input, at the frequency of 0.1 Hz. The different colors represent signals of different amplitudes set as the input, which vary in the range 3–3.5 kV with increments of 0.1 kV. In the lower part, the corresponding stroke reached by the DEA is depicted.

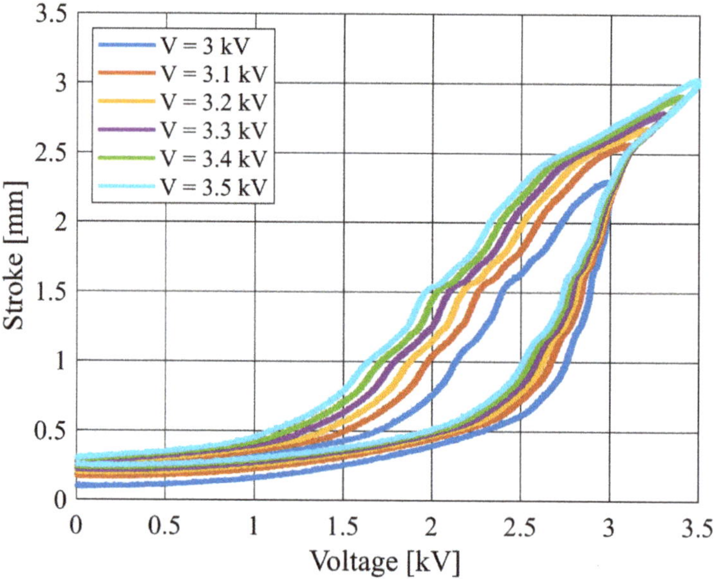

Figure 20. Stroke reached by the DEA as a function of the applied voltage, considering sinusoidal voltage inputs having frequency of 0.1 Hz and different maximum amplitudes.

6. Discussion and Future Developments

A new model-based design procedure for soft and large stroke DEAs has been presented in this work. Specifically, a novel soft actuator design is proposed based on a DE membrane combined with a polymeric biasing dome. The special mechanical characteristics of this type of biasing element allow development of large-stroke DEA systems, without introducing rigid components in the design. First, an FE model of the dome has been described, identified, and experimentally validated. Based on the obtained model, a design optimization algorithm has been developed, which allows finding of the optimal dome geometry that leads to a DEA stroke maximization. The developed procedure has finally been used to manufacture an optimized DEA prototype. The optimal geometry found for the dome allowed us to achieve a stroke of 3.1 mm, which is remarkably high for the considered type of COP-DEA.

In addition, Table 3 provides an overview of the maximum strokes, normalized over the undeformed membrane length, achieved with different bias elements investigated in the literature. The chosen references are the works by Loew et al. [34] and Hodgins et al. [29], which deal with various biasing systems based on PBS, permanent magnets, and NBS + PBS combinations. By comparing the normalized strokes obtained in these works with the value given by our polymeric dome, a remarkably higher performance can be seen. In particular, our dome outperforms the maximum stroke of the NBS + PBS by a factor higher than 1.5, making it the most effective type of NBS so far.

Table 3. Normalized stroke reached with different bias elements: positive-rate biasing spring (PBS), attracting permanent magnets (PM), negative-rate biasing spring (NBS), and novel silicone dome.

PBS	PM	NBS + PBS	Silicone Dome
0.1 [-]	0.31 [-]	0.4 [-]	0.62 [-]

Future developments will involve dynamic studies of the presented model, including identification of hysteresis occurring in the force–deformation characteristic curve, in order to improve the matching between model predictions and the experimental performance.

Moreover, the background work presented in this paper will be extended by coupling micro-layers of polymer domes with micro-layers of flexible DEA membranes, such that a matrix of cooperative actuators able to perform complex tasks can be achieved. The fully polymeric dome will be used for wearables and soft robotics applications, thanks to its low power consumption and performance that remains unchanged even in bending configuration. A new generation of cooperative micro-scale actuators, such as haptic systems, wearable loudspeakers, and microconveyors, will then be developed based one resulting DE-dome systems.

Author Contributions: Conceptualization, methodology, software, validation, formal analysis, investigation, resources, S.C.; data curation, S.C., J.N. and J.H.; writing—original draft preparation, S.C.; writing—review and editing, visualization, supervision, G.R.; project administration, funding acquisition, S.S., G.S. and G.R. All authors have read and agreed to the published version of the manuscript.

Funding: This research was funded by Deutsche Forschungsgemeinschaft (DFG, German Research Foundation), Priority Program SPP 2206 "Cooperative Multistage Multistable Microactuator Systems" (Projects: RI3030/2-1, SCHU1609/7-1, SE704/9-1).

Institutional Review Board Statement: Not applicable.

Informed Consent Statement: Not applicable.

Conflicts of Interest: The authors declare no conflict of interest. The funders had no role in the design of the study; in the collection, analyses, or interpretation of data; in the writing of the manuscript, or in the decision to publish the results.

References

1. Carpi, F.; de Rossi, D. *Dielectric Elastomers as Electromechanical Transducers*; Elsevier: Amsterdam, The Netherlands, 2008.
2. Lau, G.K.; La, T.G.; Foong, E.S.W.; Shrestha, M. Stronger multilayer acrylic dielectric elastomer actuators with silicone gel coatings. *Smart Mater. Struct.* **2016**, *25*, 125006. [CrossRef]
3. Rosset, S.; Ararom, O.A.; Schlatter, S.R.H. Shea Fabrication process of silicone-based dielectric elastomer actuators. *J. Vis. Exp.* **2016**, *2016*, 53423. [CrossRef]
4. York, A.; Dunn, J.; Seelecke, S. Experimental characterization of the hysteretic and rate-dependent electromechanical behavior of dielectric electro-active polymer actuators. *Smart Mater. Struct.* **2010**, *19*, 094014. [CrossRef]
5. Pelrine, R.E.; Kornbluh, R.D.; Joseph, J.P. Electrostriction of polymer dielectrics with compliant electrodes as a means of actuation. *Sens. Actuators Phys.* **1998**, *64*, 77–85. [CrossRef]
6. Sahu, R.K.; Saini, A.; Ahmad, D.; Patra, K.; Szpunar, J. Estimation and validation of maxwell stress of planar dielectric elastomer actuators. *J. Mech. Sci. Technol.* **2016**, *30*, 429–436. [CrossRef]
7. Hoffstadt, T.; Maas, J. Self-Sensing Control for Soft-Material Actuators Based on Dielectric Elastomers. *Front. Robot. AI* **2019**, *6*, 133. [CrossRef]
8. Rizzello, G.; Naso, D.; York, A.; Seelecke, S. A Self-Sensing Approach for Dielectric Elastomer Actuators Based on Online Estimation Algorithms. *IEEE/ASME Trans. Mechatron.* **2017**, *22*, 728–738. [CrossRef]
9. Kornbluh, R.D.; Pelrine, R.; Prahlad, H.; Wong-Foy, A.; McCoy, B.; Kim, S.; Eckerle, J.; Low, T. Dielectric elastomers: Stretching the capabilities of energy harvesting. *MRS Bull.* **2012**, *37*, 246–253. [CrossRef]
10. Hill, M.; Rizzello, G.; Seelecke, S. Development and experimental characterization of a pneumatic valve actuated by a dielectric elastomer membrane. *Smart Mater. Struct.* **2017**, *26*, 085023. [CrossRef]
11. Linnebach, P.; Rizzello, G.; Seelecke, S.; Seelecke, S. Design and validation of a dielectric elastomer membrane actuator driven pneumatic pump. *Smart Mater. Struct.* **2020**, *29*, 075021. [CrossRef]
12. Ghazali, F.A.M.; Mah, C.K.; AbuZaiter, A.; Chee, P.S.; Ali, M.S.M. Soft dielectric elastomer actuator micropump. *Sens. Actuators A Phys.* **2017**, *263*, 276–284. [CrossRef]
13. Hosoya, N.; Masuda, H.; Maeda, S. Balloon dielectric elastomer actuator speaker. *Appl. Acoust.* **2019**, *148*, 238–245. [CrossRef]
14. Moretti, G.; Rizzello, G.; Fontana, M.; Seelecke, S. A multi-domain dynamical model for cone-shaped dielectric elastomer loudspeakers. In Proceedings of the Electroactive Polymer Actuators and Devices (EAPAD) XXIII Meeting, Online, 22–26 March 2021, Volume 53. [CrossRef]
15. Qu, X.; Ma, X.; Shi, B.; Li, H.; Zheng, L.; Wang, C.; Liu, Z.; Fan, Y.; Chen, X.; Li, Z.; et al. Refreshable Braille Display System Based on Triboelectric Nanogenerator and Dielectric Elastomer. *Adv. Funct. Mater.* **2021**, *31*, 2006612. [CrossRef]
16. Qiu, Y.; Zhang, E.; Plamthottam, R.; Pei, Q. Dielectric Elastomer Artificial Muscle: Materials Innovations and Device Explorations. *Acc. Chem. Res.* **2019**, *52*, 316–325. [CrossRef] [PubMed]

17. Sîrbu, D.; Moretti, G.; Bortolotti, G.; Bolignari, M.; Diré, S.; Fambri, L.; Vertechy, R.; Fontana, M. Electrostatic bellow muscle actuators and energy harvesters that stack up. *Sci. Robot.* **2021**, *6*, eaaz5796. [CrossRef] [PubMed]
18. Carpi, F.; Frediani, G.; Gerboni, C.; Gemignani, J.; de Rossi, D. Enabling variable-stiffness hand rehabilitation orthoses with dielectric elastomer transducers. *Med. Eng. Phys.* **2014**, *36*, 205–211. [CrossRef] [PubMed]
19. Almanza, M.; Clavica, F.; Chavanne, J.; Moser, D.; Obrist, D.; Carrel, T.; Civet, Y.; Perriard, Y. Feasibility of a Dielectric Elastomer Augmented Aorta. *Adv. Sci.* **2021**, *8*, 2001974. [CrossRef] [PubMed]
20. Zhao, H.; Hussain, A.M.; Israr, A.; Vogt, D.M.; Duduta, D.M.; Clarke, D.R.; Wood, D.R. A Wearable Soft Haptic Communicator Based on Dielectric Elastomer Actuators. *Soft Robot.* **2020**, *7*, 451–461. [CrossRef]
21. Gratz-Kelly, S.; Meyer, A.; Motzki, P.; Nalbach, S.; Rizzello, G.; Seelecke, S.S. Force measurement based on dielectric elastomers for an intelligent glove providing worker assessment in the digital production. In Proceedings of the SPIE Smart Structures + Nondestructive Evaluation, Online, 27 April–9 May 2020. [CrossRef]
22. Gu, G.-Y.; Zhu, J.; Zhu, L.-M.; Zhu, X. A survey on dielectric elastomer actuators for soft robots. *Bioinspir. Biomim.* **2017**, *12*, 011003. [CrossRef]
23. Wang, L.; Hayakawa, T.; Ishikawa, M. Dielectric-elastomer-based fabrication method for varifocal microlens array. *Opt. Express* **2017**, *25*, 31708. [CrossRef]
24. Schmitt, S.; Haeufle, D. *Mechanics and Thermodynamics of Biological Muscle—A Simple Model Approach*; Springer: Berlin/Heidelberg, Germany, 2015.
25. Ghazali, F.A.M.; Jie, W.Y.; Fuaad, M.R.A.; Ali, M.S.M. Soft dielectric elastomer microactuator for robot locomotion. *Bull. Electr. Eng. Inform.* **2020**, *9*, 2286–2293. [CrossRef]
26. Berlinger, F.; Duduta, M.; Gloria, H.; Clarke, D.; Nagpal, R.; Wood, R. A Modular Dielectric Elastomer Actuator to Drive Miniature Autonomous Underwater Vehicles. In Proceedings of the 2018 IEEE International Conference on Robotics and Automation (ICRA) 2018, Brisbane, QLD, Australia, 21–25 May 2018; pp. 3429–3435. [CrossRef]
27. Prechtl, J.; Kunze, J.; Nalbach, S.; Seelecke, S.S.; Rizzello, G. Soft robotic module actuated by silicone-based rolled dielectric elastomer actuators: Modeling and simulation. In Proceedings of the ACTUATOR; International Conference and Exhibition on New Actuator Systems and Applications 2021, Online, 17–19 February 2021. [CrossRef]
28. Christianson, N.N.; Goldberg, D.D.; Deheyn, S.; Cai, M.; Tolley, T. Translucent soft robots driven by frameless fluid electrode dielectric elastomer actuators. *Sci. Robot.* **2018**, *3*, eaat1893. [CrossRef]
29. Hodgins, M.; York, A.; Seelecke, S. Experimental comparison of bias elements for out-of-plane DEAP actuator system. *Smart Mater. Struct.* **2013**, *22*, 094016. [CrossRef]
30. Hodgins, M.; York, A.; Seelecke, S. Modeling and experimental validation of a bi-stable out-of-plane DEAP actuator system. *Smart Mater. Struct.* **2011**, *20*, 094012. [CrossRef]
31. Hau, S.; Bruch, D.; Rizzello, G.; Motzki, P.; Seelecke, S. Silicone based dielectric elastomer strip actuators coupled with nonlinear biasing elements for large actuation strains. *Smart Mater. Struct.* **2018**, *27*, 074003. [CrossRef]
32. Follador, M.; Cianchetti, M.; Mazzolai, B. Design of a compact bistable mechanism based on dielectric elastomer actuators. *Meccanica* **2015**, *50*, 2741–2749. [CrossRef]
33. Bruch, P.; Loew, S.; Hau, G.; Rizzello, S.S. Fast model-based design of large stroke dielectric elastomer membrane actuators biased with pre-stressed buckled beams. In Proceedings of the SPIE Smart Structures and Materials + Nondestructive Evaluation and Health Monitoring, Denver, CO, USA, 4–8 March 2018. [CrossRef]
34. Loew, P.; Rizzello, G.; Seelecke, S. Permanent magnets as biasing mechanism for improving the performance of circular dielectric elastomer out-of-plane actuators. In Proceedings of the Electroactive Polymer Actuators Devices 2017, Portland, OR, USA, 25–29 March 2017; p. 10163. [CrossRef]
35. Berselli, G.; Vertechy, R.; Vassura, G.; Parenti-Castelli, V. Optimal synthesis of conically shaped dielectric elastomer linear actuators: Design methodology and experimental validation. *IEEE/ASME Trans. Mechatron.* **2011**, *16*, 67–79. [CrossRef]
36. Madhukar, D.; Perlitz, M.; Grigola, D.; Gai, K.J. Hsia Bistable characteristics of thick-walled axisymmetric domes. *Int. J. Solids Struct.* **2014**, *51*, 2590–2597. [CrossRef]
37. Alturki, M.; Burgueño, R. Multistable cosine-curved dome system for elastic energy dissipation. *J. Appl. Mech. Trans. ASME* **2019**, *86*, 091002. [CrossRef]
38. Neu, J.; Hubertus, J.; Croce, S.; Schultes, G.; Seelecke, S.; Rizzello, G. Fully Polymeric Domes as High-Stroke Biasing System for Soft Dielectric Elastomer Actuators. *Front. Robot. AI* **2021**, *8*, 695918. [CrossRef]
39. Jia, K.; Wang, M.; Lu, T.; Wang, T. Linear control of multi-electrode dielectric elastomer actuator with a finite element model. *Int. J. Mech. Sci.* **2019**, *159*, 441–449. [CrossRef]
40. Simone, F.; Linnebach, P.; Rizzello, G.; Seelecke, S. A finite element model of rigid body structures actuated by dielectric elastomer actuators. *Smart Mater. Struct.* **2018**, *27*, 065001. [CrossRef]
41. Croce, S.; Neu, J.; Hubertus, J.; Rizzello, G.; Seelecke, S.; Schultes, G. Modeling and simulation of compliant biasing systems for dielectric elastomer membranes based on polymeric domes. In Proceedings of the ACTUATOR; International Conference and Exhibition on New Actuator Systems and Applications 2021, Online, 17–19 February 2021; pp. 3–6.
42. Hubertus, J.; Croce, S.; Neu, J.; Rizzello, G.; Seelecke, S.; Schultes, G. Electromechanical characterization and laser structuring of Ni-based sputtered metallic compliant electrodes for DE applications. In Proceedings of the ACTUATOR; International Conference and Exhibition on New Actuator Systems and Applications 2021, Online, 17–19 February 2021; pp. 1–4.

43. Fasolt, B.; Hodgins, M.; Rizzello, G.; Seelecke, S. Effect of screen printing parameters on sensor and actuator performance of dielectric elastomer (DE) membranes. *Sens. Actuators A Phys.* **2017**, *265*, 10–19. [CrossRef]
44. Rosset, S.; Shea, H.R. Flexible and stretchable electrodes for dielectric elastomer actuators. *Appl. Phys. A Mater. Sci. Process.* **2013**, *110*, 281–307. [CrossRef]
45. Schlatter, S.; Rosset, S.; Shea, H. Inkjet printing of carbon black electrodes for dielectric elastomer actuators. *Electroact. Polym. Actuators Devices* **2017**, 1016311. [CrossRef]
46. Wacker Chemie AG ELASTOSIL Film 2030. 2013, pp. 1–4. Available online: https://www.wacker.com/h/en-us/silicone-rubber/silicone-films/elastosil-film-2030/p/000038005 (accessed on 1 July 2021).
47. Qiu, J.; Lang, J.H.; Slocum, A.H. A curved-beam bistable mechanism. *J. Microelectromech. Syst.* **2004**, *13*, 137–146. [CrossRef]
48. COMSOL. *Multiphysics Optimization Module. User's Guide*; COMSOL: Burlington, MA, USA, 2015.
49. Martins, P.A.L.S.; Jorge, R.M.N.; Ferreira, A.J.M. A comparative study of several material models for prediction of hyperelastic properties: Application to silicone-rubber and soft tissues. *Strain* **2006**, *42*, 135–147. [CrossRef]
50. Wacker Chemie AG WACKER SilGel 612 EH A/B. 2021; pp. 2–4. Available online: https://www.wacker.com/h/de-de/siliconkautschuk/silicongele/wacker-silgel-612-ab/p/000007546 (accessed on 2 December 2020).

Article

System-Level Modelling and Simulation of a Multiphysical Kick and Catch Actuator System

Arwed Schütz [1,2,*], **Sönke Maeter** [1] and **Tamara Bechtold** [1]

[1]   Jade University of Applied Sciences, Friedrich-Paffrath-Str. 101, 26389 Wilhelmshaven, Germany; soenke.maeter@student.jade-hs.de (S.M.); tamara.bechtold@jade-hs.de (T.B.)

[2]   Chair of Control Engineering, University of Augsburg, Eichleitnerstr. 30, 86159 Augsburg, Germany

*   Correspondence: arwed.schuetz@jade-hs.de

Abstract: This paper presents a system-level model of a microsystem architecture deploying co-operating microactuators. An assembly of a piezoelectric kick-actuator and an electromagnetic catch-actuator manipulates a structurally unconnected, magnetized micromirror. The absence of mechanical connections allows for large deflections and multistability. Closed-loop feedback control allows this setup to achieve high accuracy, but requires fast and precise system-level models of each component. Such models can be generated directly from large-scale finite element (FE) models via mathematical methods of model order reduction (MOR). A special challenge lies in reducing a nonlinear multiphysical FE model of a piezoelectric kick-actuator and its mechanical contact to a micromirror, which is modeled as a rigid body. We propose to separate the actuator micromirror system into two single-body systems. This step allows us to apply the contact-induced forces as inputs to each sub-system and, thus, avoid the nonlinear FE model. Rather, we have the linear model with nonlinear input, to which established linear MOR methods can be applied. Comparisons between the reference FE model and the reduced order model demonstrate the feasibility of the proposed methodology. Finally, a system-level simulation of the whole assembly, including two actuators, a micromirror and a simple control circuitry, is presented.

Keywords: finite element method; model order reduction; nonlinear; contact mechanics; multiphysics; piezoelectricity

Citation: Schütz, A.; Maeter, S.; Bechtold, T. System-Level Modelling and Simulation of a Multiphysical Kick and Catch Actuator System. *Actuators* **2021**, *10*, 279. https://doi.org/10.3390/act10110279

Academic Editors: Manfred Kohl, Stefan Seelecke and Ulrike Wallrabe

Received: 30 June 2021
Accepted: 19 October 2021
Published: 21 October 2021

Publisher's Note: MDPI stays neutral with regard to jurisdictional claims in published maps and institutional affiliations.

1. Introduction

Micromirrors and microscanners are key technologies for numerous optical applications, including laser projection [1], optical communication [2], displays [3–7], medical imaging [8,9], and light detection and ranging (LiDAR) [10–13]. This widest range of varying applications is down to their superior production costs, compact dimensions, and energy efficiency. A consequence of their popularity is the number of designs, which are commonly categorized by their actuation principle and degrees of freedom (DOFs). Actuating a micromirror typically employs either electrostatic [3,6], piezoelectric [14], electromagnetic [7,15,16], or thermoelectric [17,18] effects. These physical effects mainly differ in speed of actuation, displacement magnitude, required space and operating voltage. A micromirror's DOFs determine its number of rotational and potentially translational axes. Each design features its own maximum deflection angle and operating frequencies.

However, all these concepts share one common design aspect: the rotating mirror is mounted using structural connections as there are no ball bearings in microtechnology. Consequently, these links restrict the maximum deflection angle and cause restoring forces that the actuator has to overcome. This deficit limits the classical micromirror's use in certain applications, such as tracking, as they require large quasistatic deflections [19,20].

A novel actuator design has been proposed to overcome this limitation by omitting all structural suspensions [20]. In contrast, an unconnected hemispherical micromirror is rotated by periodic bouncing elicited by contact to an elliptically oscillating stage. This design achieves a resonant deflection angle of $\pm 35.2°$ and a maximum angular velocity of $732\,\tfrac{°}{s}$ [20]. As each contact induces rotational momentum, stepwise rotation is achieved. Up till now, challenging physics such as mechanical contact impede accurate modeling and closed-loop control, limiting the approach to a proof of concept.

The kick and catch research project [21] aims to extend this innovative system by more sophisticated actuators and highly accurate models, enabling model-based closed-loop control [22–24]. In this approach, a cooperation of a kick-actuator and a catch-actuator achieves multistable motion of the hemisphere. The former corresponds to a more sophisticated version of the stage [20], offering more control of the hemisphere's launch and landing; The latter introduces controllable forces acting on the hemisphere, thus providing access to the closed-loop control. Figure 1 illustrates the concept's basic operating principle: first, the piezoelectric kick-actuator expands. Mechanical contact transfers these sudden forces to the micromirror, launching it into a flight phase with potential rotation. During the flight, the electromagnetic catch-actuator adjusts the hemisphere's trajectory and softens its landing. Finally, the rotated micromirror lands on the kick-actuator. Repeating this sequence achieves larger rotations.

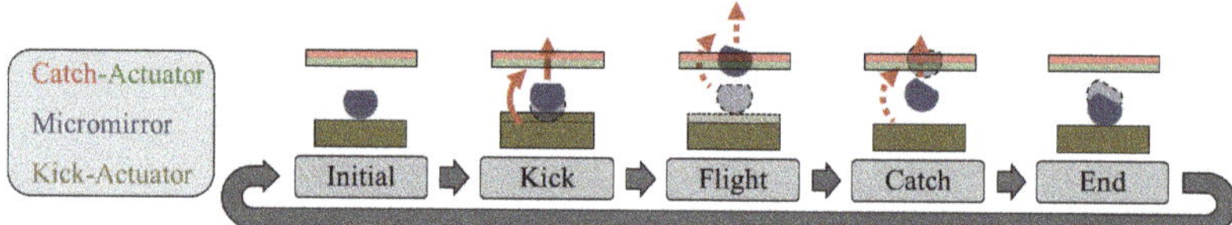

Figure 1. Working principle of the kick and catch actuator system: the kick-actuator launches the hemispherical micromirror into a flight phase. Subsequently, the electromagnetic catch-actuator controls the mirror's flight. Finally, the catch-actuator decelerates the sphere and supports its smooth landing on the kick-actuator. This sequence achieves a small rotation of the hemisphere and may be repeated periodically to achieve large deflections. Please note the symbolic nature of this illustration. Later versions of the catch-actuator may, for example, contain a three-dimensional Helmholtz-coil configuration. Furthermore, this work focuses on further developing and applying the mathematical methodology of model order reduction (MOR). For this reason, the system is simplified to vertical motion.

As optical applications require high-level accuracy, reliable system-level models are of crucial importance. This is especially true for the kick-actuator due to its mechanical contact to the hemisphere as well as its multiphysical actuation. The finite element method (FEM) allows the accurate modeling of these effects, but its computational demand prevents any system-level application.

This gap is closed by MOR methodology [25], generating system-level models of extraordinary accuracy based on, for example, finite element (FE) models. MOR reduces a model's initially high dimension by several orders of magnitude while forfeiting almost no precision. Methods to generate reduced order models (ROMs) are well established for linear systems [26], whereas mechanical contact remains challenging due to its nonlinearity and inequality constraints [27]. Another obstacle in reducing contact are mainly its three different FE implementations: Lagrange multipliers, penalty, or augmented Lagrange multipliers [28]. To enforce non-penetration, they introduce additional DOFs, penetration-opposing stiffness, or both.

General nonlinear MOR is accompanied by two particular problems: finding a reduced basis and efficiently evaluating nonlinearities [29]. The data-driven proper orthogonal decomposition (POD) [27,30–32] is the state-of-the-art method for constructing a reduced basis for a nonlinear system. Techniques [33] that do not rely on data created by extensive simulations, such as modal [34] or static derivatives, are less popular. A recent contact-specific approach separates the DOFs of the full order model (FOM) into displacements and Lagrange multipliers prior to reduction, relying on Lagrange-like contact implementation [27,35]. Evaluating nonlinearities is accompanied by extremely high computational costs and must thus be approximated to maintain the ROM's efficiency. This approximation step is known as hyper-reduction, estimating the full nonlinearity based on few local evaluations [29]. Prominent methods include the data-driven discrete empirical interpolation method (DEIM) [36] and the technique of energy conserving mesh sampling and weighting (ECSW) [27,31,37,38]. Although these methods can achieve excellent precision even for contact problems [27,31,39], they require numerous expensive FOM solutions, also referred to as snapshots. Moreover, their prediction quality is limited to cases included in their training, making data-driven ROMs less robust.

A noteworthy technique exists for reducing systems with their nonlinearities exclusively in their input, which are referred to as weakly nonlinear systems. In this case, the inputs can be grouped to reduce the number of nonlinearities [40]. The inputs of a mechanical system are external forces and grouping inputs means merging almost equal forces into a single but distributed force. The reduced number of forces allows them to be efficiently handled as nonlinear inputs for a linear ROM. Although this approach requires simple geometries, a-priori knowledge of the forces' location and mergeable forces, it offers two remarkable advantages: firstly, in contrast to state-of-the-art data-driven MOR, it does not rely on computationally expensive snapshots. Consequently, such an ROM is not restricted by its training input. Secondly, well-established methods of linear MOR, such as Krylov-subspace-generated bases [41], become available. This also includes robust methods for parametric model order reduction (pMOR), enabling the inclusion of parameters arising from material properties or geometry.

This paper applies the aforementioned reduction of weakly nonlinear systems to an impact problem on a piezoelectric kick-actuator. Hence, the method's scope is extended from contact as a boundary condition [40] to penalty-based contact between two bodies. The novel approach for this extension is to separate the kick-actuator and a sphere into individual models. The model of the kick-actuator is reduced separately and is rejoined with the sphere via contact at the system-level. This separation relocates the multibody system's nonlinearity from the stiffness matrix to the individual systems' inputs. Hence, penalty-based contact can be considered nonlinear external forces, granting access to the methodology of [40]. Further enhancement is provided through considering the expected load distribution in the FE mesh's design. As a result, the grouping-induced error is reduced compared to [40]. The final ROM complements preceding work on the electromagnetic catch-actuator [23,24] and enables system-level simulation of the entire actuator system.

The remainder of this paper is structured as follows: Section 2 provides more details on the actuator system, its FE models, their reduction by means of MOR, and a final application at system-level, demonstrating feasibility. Numerical experiments evaluate the ROM's accuracy and computational efficiency by comparing its solution to the reference. Both intermediate and final results are presented in Section 3. Finally, Section 4 offers a conclusion and identifies aspects for future investigation.

2. Materials and Methods

This section specifies the design of the kick and catch actuator system and introduces simplifications applied in this study. Subsequently, FE models of the piezoelectric kick-actuator and the electromagnetic catch-actuator are described. MOR generates system-level models deployed in a full system-level simulation. The catch-actuator has been the subject of earlier work [23,24]. Corresponding paragraphs will, therefore, be confined to summaries and updates.

2.1. Actuator System Design

The kick and catch actuator system constitutes an innovative architecture for micromirrors as it omits all structural connections. Due to this work's focus being on methods of nonlinear MOR, the system is simplified as shown in Figure 2:

- A sphere replaces the hemisphere;
- The sphere's motion is restricted to vertical displacement;
- The kick-actuator comprises a single piezoelectric chip actuator;
- All of the components are arranged concentrically.

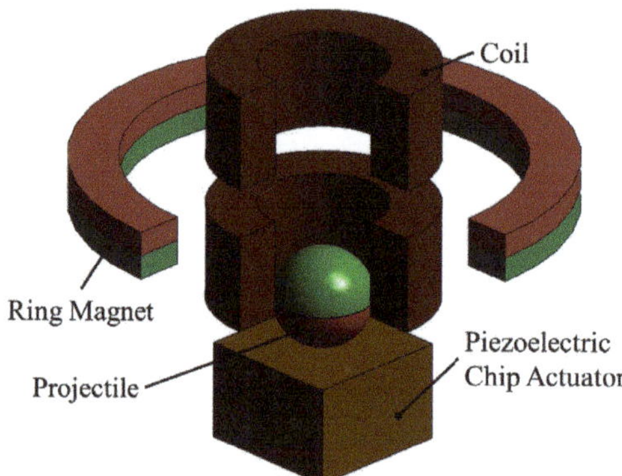

Figure 2. Sectional three-dimensional view of the simplified actuator system with labeled components. This setup deploys a piezoelectric chip actuator for kick-actuation and an assembly of two coils and a ring magnet for electromagnetic interaction. Additionally, the micromirror is included as a magnetic sphere. The setup is designed for preliminary studies. This basis will be extended by more complex assemblies to precisely manipulate the micromirror.

Even though the mirror's rotation is not modeled, these changes do not neglect physical effects and include the working principle shown in Figure 1: the expanding piezoelectric kick-actuator accelerates the sphere via mechanical contact. Subsequently, the catch-actuator controls the magnetic sphere's position via electromagnetic forces. Therefore, the setup suffices to develop methods for generating system-level models. In addition, preliminary experiments are planned using the same configuration. For that reason, all of the components were chosen from those commercially available.

For the kick-actuator, a piezoelectric PA3JEAW chip actuator by Thorlabs is deployed. A total of 33 piezoelectric layers of lead zirconate titanate (PZT)-based THP51 [42] with alternating polarization stack up to a chip actuator of $(3 \times 3 \times 2)$ mm^3. Silver electrodes in between collect or distribute electric charges. A ceramic coating prevents moisture from entering the actuator. Appendix A.2 presents all material data relevant for simulation.

The magnets employ commercial products, whereas the coils will be custom-manufactured. Table 1 denotes the cylindrical parts' dimensions and their vertical positions. Leaving a radial gap for potential guiding structures, the coils' inner radii are chosen to be 1.3 mm. The two coils' geometries are identical and deploy a copper wire with a diameter of $D_W = 25\,\mu$m. Each solenoid comprises 4074 windings, assuming a fill factor of 100%.

Table 1. Dimensions of the catch actuator's components. H corresponds to a part's height, R_o to the outer radius and R_i to the inner radius. The vertical position y refers to center of mass.

Component	H [mm]	R_o [mm]	R_i [mm]	y [mm]
Ringmagnet	1	5	4	2.5
Coil	2	2.3	1.3	1/4
Sphere	-	1	-	-

Although the setup of this work is macro-scale, a miniaturization is targeted in the long term. A corresponding design will presumably rely on following cornerstones: microcoils might be manufactured by wirebonding technology from IMTEK, Freiburg. Stacked layers of metglas allow for magnets with tailored properties. Additional mechanical leverage might extend the stroke of piezoelectric actuators.

2.2. Finite Element Models

In general, physical problems are rendered into partial differential equations (PDEs) ones that can rarely be solved analytically. The FEM spatially discretizes the computational domain, transforming a PDE into a large-scale system of ordinary differential equations (ODEs). In addition, numerical integration schemes discretize time and thus transform the system of ODEs into a system of algebraic equations. Therefore, methods of linear algebra can solve the initial physical problem. If properties of the system depend on its solution, the system is classified as nonlinear. In this case, the solution process deploys iterative schemes where each iteration solves a linearized system. For a transient analysis, these iterations arise at each time step, multiplying computational effort.

The actuator system comprises two single actuators with distinct physical domains. While the catch-actuator is an electromagnetic scenario, the kick-actuator relies on piezo-electricity and contact mechanics. As the physics of the two actuators do not effectively interfere with each other, they were investigated in separate FE analyses. All electromagnetic analyses were conducted in ANSYS Maxwell 2020 R2, all piezoelectric analyses in ANSYS Mechanical 2020 R2 and all system-level simulations in ANSYS TwinBuilder 2020 R2. A system of an AMD Ryzen 5 3600 and 16 GB RAM was employed for all analyses.

2.2.1. Electromagnetic Catch-Actuator

The catch-actuator electromagnetically controls the micromirror's position and orientation. Simplifying the device as described in Section 2.1, the mirror is exchanged with a sphere and restricted to vertical motion. Consequently, the catch-actuator interacts with the sphere via a vertical force. This electromagnetic force depends predominantly on the sphere's vertical position and the electrical currents in the solenoids. Transient effects can be ignored.

The catch-actuator's FE model exploits its rotational symmetry, allowing a two-dimensional analysis setup. An enclosing air region is introduced in addition to the geometries of the magnet, the coils and the sphere. This region represents the air between the components and in their surroundings, taking the far field into account. Its geometry in three dimensions corresponds to a cylinder of $R = 10\,\text{mm}$ and $H = 30\,\text{mm}$, centered around the actuator. All material data are given in Table A1 and assumed to be linear, reducing the complexity of a later system-level model. As mesh resolution is crucial for a FE analysis' accuracy, adaptive meshing is applied. This process convergence criteria are set to either a maximal relative error of 5×10^{-6} or 30 iterations. To compute the electromagnetic force on the sphere with respect to its dependencies, both coil currents and the sphere's position are parameterized. The sphere's vertical position ranges from $-2.5\,\text{mm}$ to $7.5\,\text{mm}$ in 41 steps, the coil currents from $-0.1\,\text{A}$ to $0.1\,\text{A}$ in increments of $0.1\,\text{A}$. Finally, the catch-actuator is analyzed in a set of 369 static electromagnetic simulations. A more extensive description is available in [23,24].

2.2.2. Piezoelectric Kick-Actuator

The kick-actuator launches the hemispherical mirror by a sudden transfer of momentum and also serves as a landing platform. In the context of this work, the PA3JEAW actuator launches a rigid sphere in response to an electrical excitation. Although simplified, this sequence features transient multiphysics and nonlinearities due to piezoelectricity and mechanical contact.

The modeled geometry represents the piezoelectric chip actuator defined in Section 2.1. Both the ceramic coating and the electrodes have little influence on the actuator's mechanical behavior. Therefore, their geometry is excluded from the model. All electrodes are considered ideally conductive, allowing for coupling electrical DOFs of all anodes and cathodes respectively. Moreover, the cathode is grounded and electric loads may be applied to the anode. The 33 piezoelectric layers of alternating polarization are modeled as one block of THP51. A mapped layered mesh reintroduces their structure and Appendix A.2 provides the THP51's material data. The sphere is considered rigid to reduce complexity of nonlinear iterations. Thus, only the sphere's density of $\rho = 7850\,\frac{\mathrm{kg}}{\mathrm{m}^3}$ remains relevant. In addition, several symmetries are introduced by restricting the sphere to vertical motion. As a result, modeling a quarter of the setup as shown in Figure 3a suffices, significantly saving computational resources. This is particularly useful for transient nonlinear analyses. Furthermore, any vertical displacement of the bottom face is set to zero, representing an adhesive connection. Additional displacement constraints prevent rigid body motion of the chip actuator. Simulating initial rigid body motion in impact setups brings no insights at full computational costs for each time step. To minimize this issue, the sphere is modeled 1 μm above the actuator's top surface and assigned an initial velocity. The transient analysis deploys an implicit Newmark time integration with a time-step of 50 ns to an end time of 15 μs.

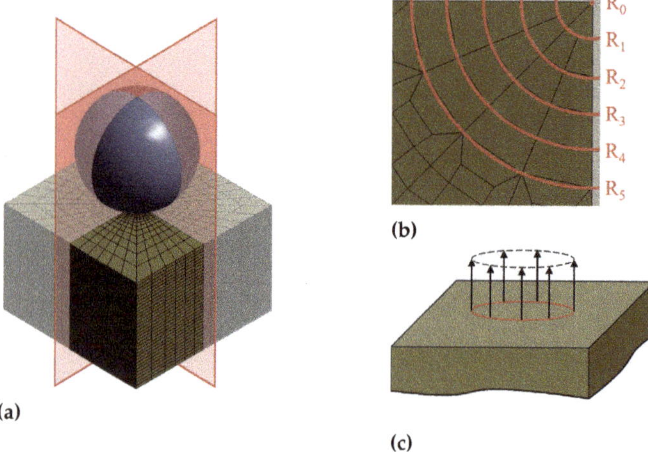

Figure 3. Considerations for the FEM model of the piezoelectric kick-actuator: (**a**) Symmetry allows to simulate only one quarter of the model, saving computational effort. (**b**) A mapped mesh of concentric circles around the location of contact results in equal vertical nodal forces per ring. The center node and the five rings are enumerated from R_0 to R_5. (**c**) Contact-induced forces on the kick-actuator are equal per ring. Black arrows indicate the vertical force distribution for a single ring. Separately modelling the sphere and the kick-actuator provides access to the MOR methodology proposed in [40].

The chip actuator is spatially discretized into a linear mapped FE mesh for three reasons: firstly, a mapped mesh allows to accurately model the piezoelectric layers. Secondly, a mapped instead of a free mesh avoids numerically induced asymmetries in the contact forces. Thirdly, and most importantly, this part of the mesh is the basis for efficiently reducing contact. The procedure in [40] groups together nodes of equal external forces. Hence, it is crucially important to have a mesh that supports such clustering. As a sphere causes the contact forces, locations on a concentric circle experience effectively equal vertical forces. No radial forces occur due to settings chosen for contact modeling. Therefore, meshing the contact area in five concentric rings as illustrated in Figure 3b is an appropriate choice. The rings have a radial distance of 50 μm as a fine mesh in contact regions improves accuracy and nonlinear convergence.

Excluding contact and the sphere for now, the kick-actuator constitutes a linear piezoelectric problem. Employing a strong coupling, such a system is governed by a system Σ_P of ODEs expressed as:

$$\Sigma_P = \begin{cases} \underbrace{\begin{bmatrix} M_{xx} & 0 \\ 0 & 0 \end{bmatrix}}_{M} \underbrace{\begin{bmatrix} \ddot{x}^* \\ \ddot{V} \end{bmatrix}}_{\ddot{x}} + \underbrace{\begin{bmatrix} E_{xx} & 0 \\ 0 & 0 \end{bmatrix}}_{E} \underbrace{\begin{bmatrix} \dot{x}^* \\ \dot{V} \end{bmatrix}}_{\dot{x}} + \underbrace{\begin{bmatrix} K_{xx} & K_{xV} \\ K_{Vx} & K_{VV} \end{bmatrix}}_{K} \underbrace{\begin{bmatrix} x^* \\ V \end{bmatrix}}_{x} = f = \underbrace{\begin{bmatrix} B_x \\ B_V \end{bmatrix}}_{B} u \\ y = \underbrace{\begin{bmatrix} C_x & C_V \end{bmatrix}}_{C} \begin{bmatrix} x^* \\ V \end{bmatrix}, \end{cases} \tag{1}$$

where $M, E, K \in \mathbb{R}^{n \times n}$ are the mass, damping and stiffness matrices. The blockmatrices' subscripts indicate their physical domain: "$_{xx}$" corresponds to structural mechanics, "$_{VV}$" to electrics and "$_{xV}$"/"$_{Vx}$" to their coupling. Consequently, the state vector $x \in \mathbb{R}^n$ incorporates n_x nodal displacements $x^* \in \mathbb{R}^{n_x}$ and n_V nodal electrical potentials $V \in \mathbb{R}^{n_V}$. Further, the input vector is denoted by $u \in \mathbb{R}^p$, the user-defined output vector by $y \in \mathbb{R}^q$. The matrix $B \subset \mathbb{R}^{n \times p}$ distributes the inputs, whereas $C \in \mathbb{R}^{q \times n}$ computes the outputs. From a physical perspective, the columns of B contain "force shapes" that are scaled by u. Thus, their product Bu equals a vector of total external forces $f \in \mathbb{R}^n$. When using Rayleigh damping, the mechanical damping matrix $E_{xx} \in \mathbb{R}^{n \times n}$ is given as a weighted sum of mass and stiffness matrix. With constant Rayleigh coefficients α and β, the relation is expressed by:

$$E_{xx} = \alpha \, M_{xx} + \beta \, K_{xx}. \tag{2}$$

Contact in FE analyses adds two steps to the solution process [27]: first, the solver checks for contact. If contact is detected, non-penetration is enforced via nonlinear iterations. Three methods have been established to handle this constraint: Lagrange multipliers, penalty and augmented Lagrange multipliers. This work deploys the penalty method which introduces virtual springs opposing penetration. Their force depends linearly on the penetrating volume or distance and vanishes as soon as the bodies leave contact. Although this method tolerates small penetrations and may cause ill-conditioning, it offers superior convergence and does not introduce new DOF [43]. A constant contact stiffness of 100,000 $\frac{\text{N}}{\text{mm}}$ has been chosen. The setup is assumed to be frictionless, as friction barely contributes due to the model's simplifications. This assumption might not yield accurate results for more complex scenarios. Further, adhesion is neglected as corresponding physical effects do not contribute in relevant orders of magnitude.

Contact occurs if the distance $d(x) \in \mathbb{R}^{N_C}$ between certain nodes falls below zero. Mathematically, this condition is denoted by the following inequality constraints [39,44]:

$$d(x) = B_C x + g_0 \geq 0, \tag{3}$$

where each row represents one of the N_C inequality constraints. The matrix $B_C \in \mathbb{R}^{N_C \times n}$ selects nodal displacements for each constraint and $g_0 \in \mathbb{R}^{N_C}$ denotes initial gap sizes. Since x is unknown, the contact status can only be determined upon solution. Therefore, the contact is adjusted in iterative solutions until convergence. If $d(x) < 0$, penetration occurs and Equation (3) is violated. In this case, the penalty algorithm introduces counteracting forces $f_{Penalty}(x)$, approximately restoring $d(x) \geq 0$. Only violated constraints are enforced and thus contact forces are applied to a corresponding subset N_C^* of all N_C constraints. These forces linearly depend on penetration via a penalty stiffness ε and read as:

$$f_{Penalty}(x) = \begin{cases} \varepsilon\, B_{C,v}^T\, d_v(x) = \underbrace{\varepsilon\, B_{C,v}^T\, B_{C,v}\, x}_{K_C(x)} + \underbrace{\varepsilon\, B_{C,v}^T\, g_{0,v}}_{f_C(x)} & \text{if violated} \\ 0 & \text{else.} \end{cases} \tag{4}$$

Here, the subscript "$_v$" denotes subsets of respective quantities that exclusively contain violated constraints. Since detecting constraint violations requires the solution, the subsets depend on x and may change for each iteration. Rewriting the forces as contact stiffness $K_C \in \mathbb{R}^{n \times n}$ and contact force $f_C \in \mathbb{R}^n$ emphasizes their physical interpretation as additional stiffness. Their nonlinearity is indicated as a function of x, replacing their subscript "$_v$". Combining Equations (1) and (4) provides the system Σ_{PC} of piezoelectric contact as:

$$\Sigma_{PC} = \begin{cases} M\ddot{x} + E\dot{x} + (K + K_C(x))x = f - f_C(x) \\ y = Cx. \end{cases} \tag{5}$$

Contact analyses usually feature multiple bodies that do not interact until they are in contact. Therefore, all system matrices are block diagonal until contact introduces $K_C(x)$, coupling the bodies. In the case of a rigid body, its blocks in the system matrices are of dimension $\mathbb{R}^{1 \times 1}$.

2.3. Model Order Reduction

MOR approximates a large-scale system by an ROM of a drastically smaller dimension. The ROM hardly forfeits any accuracy while providing a speed-up of several orders of magnitude. FE solvers are a natural application for MOR, as they assemble and solve large-scale systems of ODEs. Figure 4 illustrates this idea for the piezoelectric kick-actuator.

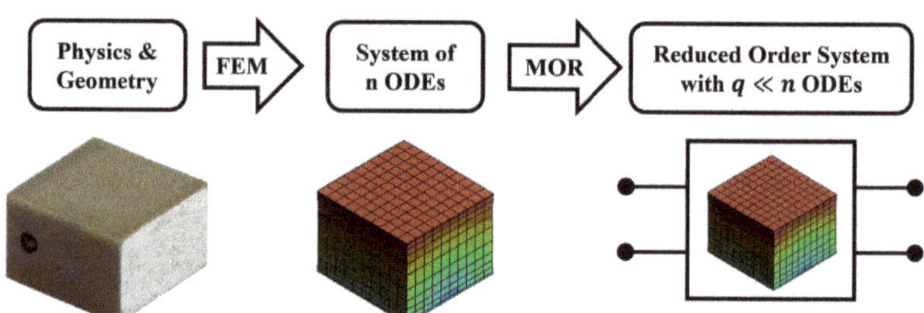

Figure 4. Schematic workflow of MOR, illustrated for the Thorlabs PA3JEAW piezoelectric chip actuator: first, a physical problem to be investigated is chosen and subsequently modeled based on the FEM. The FEM spatially discretizes the computational domain, creating a large set of ODEs. MOR projects these ODEs onto a low-dimensional subspace, reducing the number of equations by several orders of magnitude. Finally, the resulting ROM is ready to use for commercial system-level simulation software. The picture of the actuator is adapted from [45] with the friendly permission of Thorlabs GmbH.

One robust and accurate method for linear models is projecting a large-scale system onto a low-dimensional Krylov subspace [41,46]. For this approach, the state vector x is expressed as:

$$x = V\, x_r + \varepsilon, \tag{6}$$

where $V \in \mathbb{R}^{n \times q}$ is the orthogonal basis of the corresponding Krylov subspace and $x_r \in \mathbb{R}^q$ is the reduced state vector. The approximation error $\varepsilon \in \mathbb{R}^n$ is cut off and contains the part of x that does not lie in the Krylov subspace. The subspace's basis vectors may be orthonormalized by the block Arnoldi algorithm. Its dimension q is several orders of magnitude smaller than n, that is, $q \ll n$. Projecting the linear second order system Σ_P in Equation (1) onto V, one obtains the reduced model:

$$
\Sigma_{Pr} =
\begin{cases}
\underbrace{V^T M V}_{M_r}\, \ddot{x}_r + \underbrace{V^T E V}_{E_r}\, \dot{x}_r + \underbrace{V^T K V}_{K_r}\, x_r = \underbrace{V^T B}_{B_r}\, u \\
y = \underbrace{C V}_{C_r}\, x_r.
\end{cases}
\tag{7}
$$

The projection significantly reduces the system's dimension to $x_r \in \mathbb{R}^r$, $M_r, E_r, K_r \in \mathbb{R}^{r \times r}$, $B_r \in \mathbb{R}^{r \times p}$ and $C_r \in \mathbb{R}^{q \times r}$. In addition, the reduction preserves all inputs u and outputs y. However, the projected system lacks physical meaning.

MOR expresses the state vector x as a linear combination of predefined "shapes" (columns of V). Further, the reduced state vector corresponds to the weights of each shape. From a mathematical perspective, Krylov subspace-based MOR matches the Taylor expansions of the ROM's and the FOM's transfer functions around a frequency s_0 for the first q moments.

However, such reduction of multiphysical systems does not necessarily preserve its stability [47–49]. Therefore, Schur after MOR [48] is applied to reintroduce stability to the piezoelectric system's ROM. Translating the reduced system into very high speed integrated circuit hardware description language (VHDL) creates a convenient interface to commercial system-level simulation software.

As mentioned in Section 1, additional challenges accompany nonlinear MOR as a nonlinear system cannot be directly reduced. If the nonlinearities can be moved to the input, methods of linear MOR can be applied. In this case, the ROM evaluates the same nonlinear forces as the FOM. However, efficiency decreases with each nonlinear evaluation. If some nonlinear forces are almost the same, they can be approximated by computing a single force and distributing it to relevant nodes [40]. Therefore, the number of nonlinearities is reduced and the ROM gains efficiency. Obviously, the grouping process introduces errors depending on how similar the forces are. An appropriately designed mesh diminishes the error and also supports the choice of which forces to group.

However, penalty-based contact not only introduces nonlinear external forces, but also a nonlinear stiffness matrix as shown in Equation (5). The method of [40] only applies to the former, but cannot handle the latter. Compatibility can be achieved by separately modeling and reducing the bodies. The separation recasts the nonlinear coupling within the stiffness matrix of the multibody system into external nonlinear forces of the separated systems. Subsequently, these forces can be grouped for each body as proposed in [40]. Since the sphere is considered rigid and is restricted to vertical motion, it can be represented by a point mass. Moreover, its dynamics only depend on the total vertical contact force.

Therefore, only contact forces acting on the kick-actuator must be grouped. Because these forces result from contact to a sphere, circles around the center of impact each have almost identical vertical forces, as shown in Figure 3c. In addition, a FE mesh of a center node and five concentric rings as displayed in Figure 3b minimizes errors and facilitates the grouping process. The vertical forces are grouped for each ring, resulting in a total of six penetration-dependent forces. To compute the penetrations, the system's outputs must include all the rings' vertical displacements. Furthermore, the kick-actuator also requires the electric charge on the actuator's coupled anodes Q_{Anode} as an input and their voltage V_{Anode} as an output. The resulting weakly nonlinear system Σ_{KA} of the kick-actuator is given by:

$$
\Sigma_{KA} =
\begin{cases}
M \ddot{x} + E \dot{x} + K x = B\, u(x, u_{Y,S})) \\
y = C x
\end{cases}
,
\tag{8}
$$

$$u(x, u_{Y,S}) = \begin{bmatrix} F_{C,R_0}(u_{Y,R_0}, u_{Y,S}) \\ \vdots \\ F_{C,R_5}(u_{Y,R_5}, u_{Y,S}) \\ Q_{Anode} \end{bmatrix}, \quad y = \begin{bmatrix} u_{Y,R_0} \\ \vdots \\ u_{Y,R_5} \\ V_{Anode} \end{bmatrix}. \tag{9}$$

The differences to the general piezoelectric system in Equation (1) are the system's inputs and outputs. The output vector y includes all rings' vertical displacements u_{Y,R_i} as they are required to compute the penetrations. The output matrix B includes six normalized ring-based force shapes as illustrated in Figure 3c and the electric charge distribution on the anodes. Accordingly, the nonlinear input vector $u(x, u_{Y,S})$ scales each force shape by the corresponding ring's total contact force $F_{C,R_i}(x, u_{Y,S})$. For a ring R_i, this force depends on the kick-actuator's deformation x and the position of the sphere $u_{Y,S}$; it is denoted by:

$$F_{C,R_i}(u_{Y,R_i}, u_{Y,S}) = \begin{cases} \varepsilon\left(u_{Y,S} - u_{Y,R_i} + g_{0,R_i}\right) & \text{for penetration at } R_i \\ 0 & \text{else,} \end{cases} \tag{10}$$

where u_{Y,R_i} is the vertical displacement of ring R_i and g_{0,R_i} is the initial vertical gap between the ring and the sphere.

2.4. System-Level Simulation

This section deploys a full system-level simulation of the simplified kick and catch actuator system, demonstrating feasibility. System-level simulation is highly beneficial for several reasons: first, the simulations are multiple orders of magnitude faster than at device-level with comparable accuracy. This speed-up allows us to efficiently study different load cases. Further, sub-models from different physical domains can be combined. In addition, co-simulation with driving/control circuitry becomes feasible, thus improving the system's development as a whole. Although the output quantities have to be chosen beforehand, every quantity involved in the calculation can be promoted to an output.

In order to model the actuator system, the kick-actuator's ROM and the sphere's point mass are coupled via six lumped contact springs. Each spring represents a single ring's vertical contact. If a spring detects penetration, it applies a corresponding force to the two bodies. For the kick-actuator's ROM, each spring force scales its corresponding force shape as indicated in Figure 3c.

After contact is reestablished, an equivalent circuit (EC) of the electromagnetic catch-actuator complements the actuator system. An EC is a lookup table based on parametric simulation and is an industrial standard method. Here, the EC provides the electromagnetic force on the sphere with respect to its vertical position and the two coil currents. Section 2.2.1 describes each parameter's range and resolution.

A PID-based closed-loop control of the kicked sphere's position constitutes a representative application. The global coordinate system's origin is located at the kick actuator's top surface and, hence, the sphere starts at 1 mm. To launch the sphere, a heuristically determined rectangular pulse voltage of 100 V and 25 μs is applied. The control aims to catch the sphere at a set point of 2.5 mm. Although the system is nonlinear and requires sophisticated control strategies, a PID is deployed for the sake of demonstration. This controller applies the same current to both coils but negates the lower solenoid's input. The controller's three gains are determined via optimization, aiming for little overshoot and settling time. They are given by $K_P = 50.000$, $K_I = 10$, and $K_D = 20$. Figure 5 provides the complete schematic diagram of the controlled kick and catch actuator system.

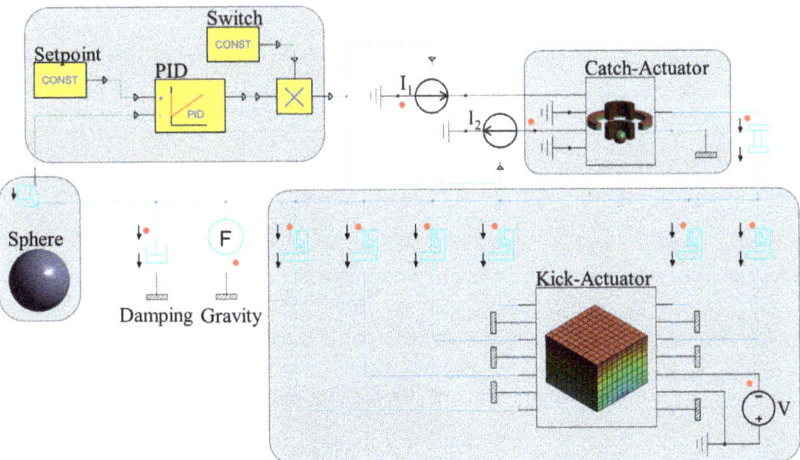

Figure 5. Schematic diagram of the kick and catch actuator system at system-level, extended by a PID-based position control of the sphere. The four grey areas indicate the system's major components: the controller, the electromagnetic catch-actuator, the piezoelectric kick-actuator, and the sphere (clockwise from top left). The catch-actuator is modelled as an equivalent circuit, the kick-actuator as an ROM and the micromirror as a point mass.

3. Results

The following section presents the results corresponding to the parts of Section 2. Firstly, the electromagnetic force that the catch-actuator applies to the magnetic sphere is evaluated. As this work focuses on reducing the piezoelectric kick-actuator, its ROM will be evaluated with respect to a FE reference solution. This evaluation comprises multiple aspects and assesses both accuracy and computational efficiency. These aspects are harmonic analyses of the linear kick-actuator; the static contact force with respect to the sphere's imprint depth; and a transient impact. A review of the system-level models' performance in an application completes the section.

3.1. Electromagnetic Force

The catch-actuator applies an electromagnetic force onto the sphere. This force depends on the solenoids' electrical currents and the sphere's position. A parametric analysis evaluates the combinations specified in Section 2.2.1. These results can be transformed into an EC of the catch-actuator, which can be deployed at system level. Figure 6 plots the vertical force with respect to the sphere's position. Each line corresponds to a distinct combination of coil currents. Consequently, the combination of zero currents shows solely the ring magnet's force. Each zero of any of the functions would correspond to an equilibrium position if there were no other forces. A negative or positive gradient at such an equilibrium position would determine whether the position is stable or unstable respectively.

3.2. Reduced Order Model of the Kick-Actuator

The kick-actuator's ROM forms the centerpiece of this work and reduces the original $n = 16{,}702$ equations to $q = 30$. Its performance is assessed in three numerical experiments. Firstly, the harmonic responses of the purely piezoelectric ROM are compared to the FOM's solutions. Another numerical experiment investigates the ROM's prediction quality regarding the static contact force. Finally, a transient impact is simulated with the FOM and the ROM. The sphere's trajectory and the contact force are tracked and compared. All three experiments report both the ROM's relative error and the observed speedup.

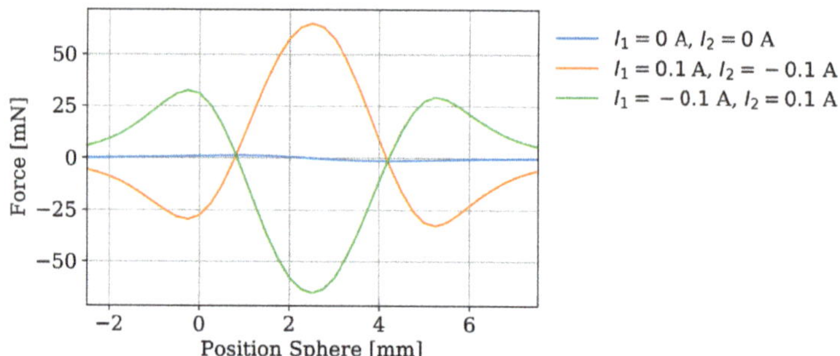

Figure 6. The vertical electromagnetic force acting on the spherical magnet, plotted over its position. Each combination of coil currents is shown as a single line. The catch-actuator's vertical symmetry induces a symmetric force for equal but opposed currents.

3.2.1. Harmonic Response

A well-established method to evaluate a linear ROM's accuracy is its harmonic responses. Based on these responses and reference solutions of the FOM, relative errors can be computed. A separate analysis is deployed for each input with a single unit load for that input. Hence, the number of analyses equals the number of inputs. For each analysis, all outputs need to be considered. Therefore, the total number of harmonic results is given by the product of the numbers of inputs and outputs. Since the kick-actuator's ROM features seven inputs and as many outputs; only a selection is shown here, while more extensive results are presented in Appendix B. Figure 7a shows a representative harmonic response for a frequency range of 0 kHz to 500 kHz in 200 steps. This wide frequency range is chosen to include the peak at about 250 kHz. Figure 7b complements the first graph as it plots the ROM's relative errors for all seven outputs. The ROM achieves excellent accuracy for small frequencies since it approximates the original transfer function around 0 Hz. Hence, accuracy at higher frequencies can be improved easily by extending the reduced basis with basis vectors for higher frequencies. The FOM's solution takes 327 s on the hardware specified in Section 2.2, whereas the ROM provides the solution after 45.0 ms.

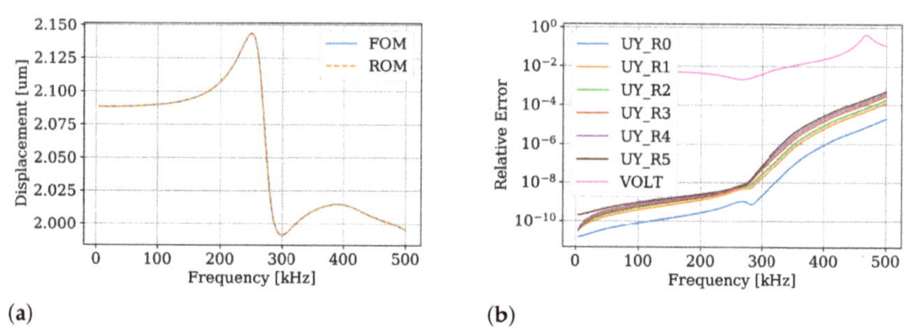

Figure 7. (**a**) Comparison of the center node's harmonic displacement amplitude obtained by the original FEM model and the ROM in a frequency range of 0 kHz to 500 kHz. (**b**) The harmonic relative error of the ROM's solution for all seven outputs, demonstrating its accuracy. The ROM approximates the original transfer function at an expansion point of 0 Hz. Consequently, all errors are the lowest at this frequency and increase with higher frequencies. Extending the reduced basis with vectors for higher frequencies enhances accuracy if needed. The error plots for the remaining six inputs are provided by Appendix B.

3.2.2. Contact Force

Penalty-based contact couples bodies within the system's stiffness matrix. Thus, a static analysis suffices to assess contact. The bodies are adjusted to touch and a subsequent displacement of $-50\,\mu\mathrm{m}$ in 250 steps triggers the contact forces. The contact forces of both FOM and ROM are plotted in Figure 8a over the sphere's imprint depth. Figure 8b presents the ROM's relative error with respect to the FE reference solution for contact between sphere and kick-actuator. The error's magnitude emphasizes the prediction quality and hence supports the methodology proposed in this study. The sudden changes in the error occur when the next ring is in contact and starts contributing to the force. Solving a single step for the FOM takes 8.206 s. The ROM provides results of almost the same accuracy within 21.96 ms per single solution.

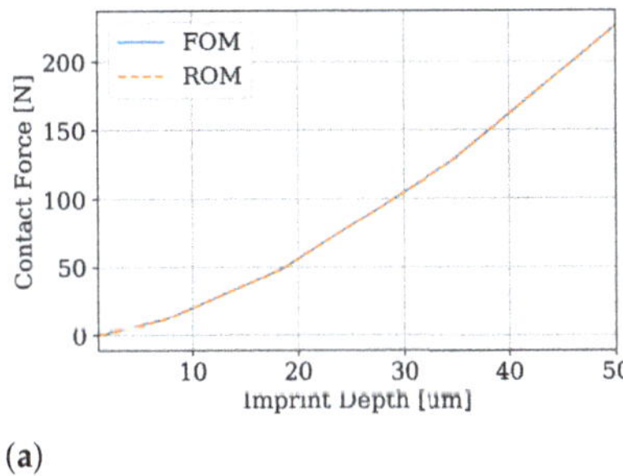
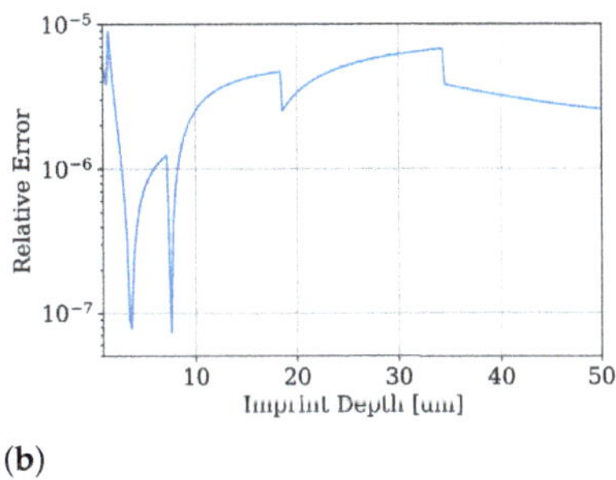

(a)   (b)

Figure 8. Static force opposing the sphere's displacement into the kick-actuator. The sphere starts just in contact with the actuator's top surface and is displaced 50 µm into the surface in increments of $-0.2\,\mu\mathrm{m}$. (**a**) Solutions of the reference FEM model and the ROM. Note that the force is one quarter of the full contact force. (**b**) The ROM's relative error, demonstrating its accuracy.

3.2.3. Transient Impact

In addition to previous experiments, a transient impact analysis provides information about the error's evolution in time. To minimize initial rigid body motion, the sphere starts 1 µm above the kick-actuator's top surface with an initial velocity of $-1\,\frac{\mathrm{m}}{\mathrm{s}}$. A duration of 15 µs is simulated with a constant time increment of 50 ns. However, the FE simulation deploys a Newmark time integration method that is not available at system-level, potentially causing deviations in transient analyses.

Figure 9 contains two plots: Figure 9a compares the two model's solutions for the sphere's trajectory during impact; and Figure 9b provides the ROM's relative error over time. Although errors accumulate with time in a transient analysis, the ROM provides accurate results. The main reason for an increasing error is the different time integration scheme deployed at system-level. After circa 11 µs, the sphere leaves contact, inducing a high error. Supporting previous statements on computational efficiency, the ROM is solved after 5.983 s and thus more than 1750 times faster than its reference that takes 10,675 s. This noteworthy gain in efficiency allows to refine the time discretization at system-level to enhance its accuracy if needed.

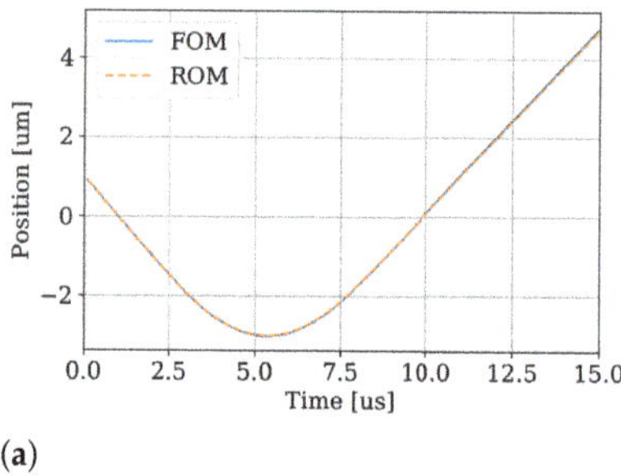 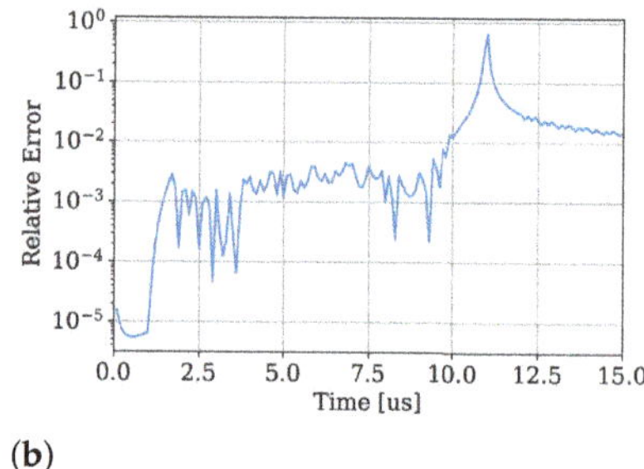

(a)   (b)

Figure 9. The sphere's vertical position during impact measured from its lowest point. (**a**) The solutions of the reference FE model and the ROM. (**b**) The ROM's relative error increases in time as deviations accumulate. The sphere leaves contact after 11 µs, causing a high error at this point in time. Note that, due to limitations of commercial software, different methods are used for time integration.

3.3. System-Level Simulation

Finally, an application at system-level combines all components of the kick and catch actuator system. As described in Section 2.4, a closed-loop position control of the kicked sphere is considered. Commercial FE software cannot simulate such a multiphysical, controlled system with justifiable effort. As a result, the setup is investigated exclusively at the system-level.

Figure 10a presents the sphere's trajectory. The catch-actuator stabilizes the sphere at the desired setpoint of 2.5 mm after 3 ms. In addition, Figure 10b shows the vertical displacement of the kick-actuator's center node. Since the accuracy of all individual system-level models is demonstrated in preceding experiments, similar reliability features in this entire setup, too. Simulating this 5 ms sequence with a constant time-step of 10 µs requires 515 ms of computation.

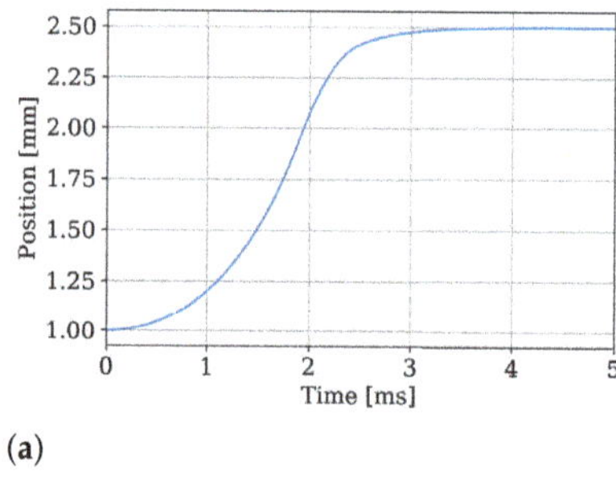 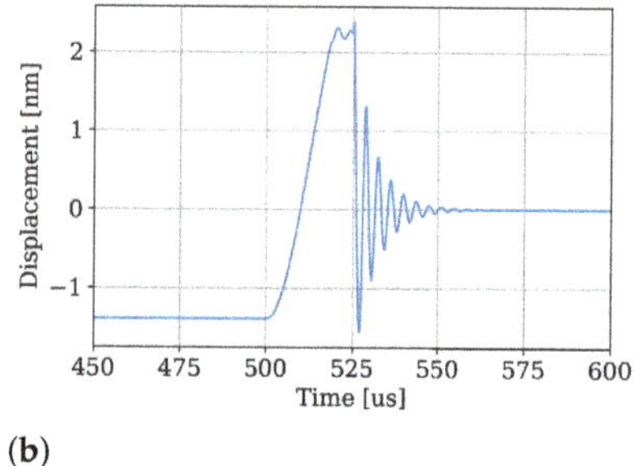

(a)   (b)

Figure 10. The kick and catch actuation at system-level. (**a**) The sphere's vertical position over time for the full duration of 5 ms, showing a clear catch at 2.5 mm. (**b**) Vertical position of the center node for 450 µs to 600 µs, illustrating the kick actuation.

4. Discussion and Outlook

This work presents a system-level model of cooperating kick and catch microactuators. The key contribution of this work is creating a FE-based system-level model of the kick-actuator including its mechanical contact to a spherical micromirror. By separately analyzing the contact bodies, their nonlinear coupling within the stiffness matrix can be transferred into an input nonlinearity. This modification leads to a linear FE model with nonlinear input including the penalty-based contact. Therefore, methods of linear MOR and techniques to reduce the number of nonlinearities [40] can be employed. A further enhancement includes considering the distribution of nonlinear forces in the spatial FE discretization, which improves accuracy. The resulting reduced order model achieves excellent accuracy and decreases the CPU time of transient simulation by several orders of magnitude. In the static comparison in Section 3.2.2, the ROM has a maximal relative error of less than 10^{-5} with respect to the original finite element model. Such efficient and accurate ROM can be used for optimization of control circuitry. Please note that this methodology is not limited to the kick-actuator considered here, but will perform equally well on similar contact scenarios.

However, our method requires an appropriately designed mesh in the location of impact. Hence, the location must be known in advance and remain constant during impact. Parametric model order reduction would allow dynamical repositioning of the impact location. Further, this work makes use of a rigid body of simple shape. Flexible bodies may be included, but require matching meshes in the contact region. More complex geometries can be handled, but may require preceding simulations to determine the distribution of contact forces. Aside from that, the kick-actuator's contact only covers imprints of limited depth. This depth can be arbitrarily chosen and deeper impacts would leave the scope of linear elasticity. Therefore, the depth is more a parameter of the model than a limitation. A potential challenge for the proposed method is friction: while Coulomb friction can be introduced easily, more sophisticated friction models remain challenging. In conclusion, this approach is highly advantageous, but is limited to problems of a certain structure.

Future research will aim to extend the method's scope to more general problems. A combination with parametric MOR seems promising, especially since it is well-suited to linear systems. In the context of the kick-actuator, this extension might allow an unrestricted movement of the sphere. Another goal is to apply the proposed methodology to a different design of the kick-actuator, featuring electrostatic pull-in similar to [20]. Furthermore, experimental investigations are ongoing and comparisons to simulations are planned. The design of the actuator will gradually become more complex to achieve the targeted motion cycle.

Author Contributions: Conceptualization, A.S. and T.B.; methodology, A.S.; software, A.S.; investigation, A.S. and S.M.; writing—original draft preparation, A.S.; writing—review and editing, A.S., S.M., and T.B.; visualization, A.S.; supervision, T.B.; funding acquisition, T.B. All authors have read and agreed to the published version of the manuscript.

Funding: This research was funded by Deutsche Forschungsgemeinschaft (German Research Foundation) grant number 424616052.

Institutional Review Board Statement: Not applicable.

Informed Consent Statement: Not applicable.

Data Availability Statement: The data can be provided by the author A.S. upon reasonable request.

Acknowledgments: We thank Thorlabs GmbH for supplying additional information and their kind permission to use pictures from their website.

Conflicts of Interest: The authors declare no conflict of interest. The funders had no role in the design of the study; in the collection, analyses, or interpretation of data; in the writing of the manuscript, or in the decision to publish the results.

Abbreviations

The following abbreviations are used in this manuscript:

DEIM	discrete empirical interpolation method
DOF	degree of freedom
EC	equivalent circuit
ECSW	energy conserving mesh sampling and weighting
FE	finite element
FEM	finite element method
FOM	full order model
LiDAR	light detection and ranging
MOR	model order reduction
ODE	ordinary differential equation
PDE	partial differential equation
pMOR	parametric model order reduction
POD	proper orthogonal decomposition
PZT	lead zirconate titanate
ROM	reduced order model

Appendix A. Material Data

Appendix A.1. Electromagnetic Material Properties

Table A1. Data for the linear materials used in the FE model of the catch actuator. μ_{rel} is the relative permeability, B_r the remanence, and H_c the coercivity.

Material	μ_{rel} [-]	B_r [T]	H_c [$\frac{A}{m}$]	Component
Ferrite	1.0611	0.2	150,000	Ring magnet, sphere
Copper [50]	1	-	-	Coils
Air [50]	1	-	-	Region

Appendix A.2. THP51

The numerical experiments in study deployed the piezoelectric chip actuator PA3JEAW by Thorlabs GmbH. This actuator is made of THP51, a soft piezoelectric ceramic based on PZT [42]. THP51 is an orthotropic material and has a density of $\rho = 7700 \ \frac{kg}{m^3}$. Its stiffness matrix C_E for no or constant electric field is given by

$$C_E = \begin{bmatrix} 158.9 & & & & & \\ 113.0 & 158.9 & & & \text{sym.} & \\ 117.1 & 117.1 & 144.2 & & & \\ 0 & 0 & 0 & 14.08 & & \\ 0 & 0 & 0 & 0 & 14.08 & \\ 0 & 0 & 0 & 0 & 0 & 22.94 \end{bmatrix} \text{GPa,} \tag{A1}$$

Its piezoelectric coupling matrix in stress-charge form e by

$$e = \begin{bmatrix} 0 & 0 & 0 & 0 & 13.38 & 0 \\ 0 & 0 & 0 & 13.38 & 0 & 0 \\ -3.914 & -3.914 & 27.50 & 0 & 0 & 0 \end{bmatrix} \frac{C}{m^2}, \tag{A2}$$

And its relative electric permittivity $\frac{\varepsilon_T}{\varepsilon}$ by

$$\frac{\varepsilon_T}{\varepsilon} = \begin{bmatrix} 3370 & 0 & 0 \\ 0 & 3370 & 0 \\ 0 & 0 & 3300 \end{bmatrix}. \tag{A3}$$

All material data represent average values and may deviate slightly due to their manufacturing process.

Appendix B. Additional Harmonic Evaluations of the ROM

The system of the linear kick-actuator without contact includes seven inputs and seven outputs. Therefore, to assess the ROM's accuracy with respect to the FOM's solutions, each output has to be evaluated for each input. The relative errors of all outputs for a unit load in the first input are shown in Figure 7b. The following six plots present the relative errors of all outputs for each of the remaining inputs.

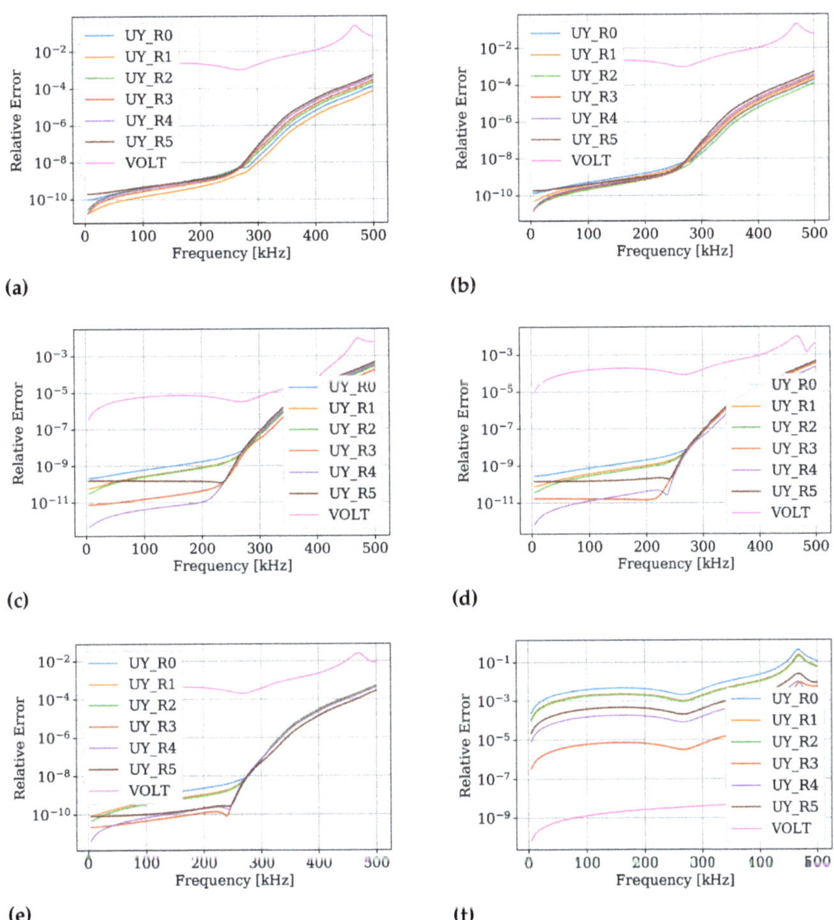

Figure A1. Harmonic relative error of each of the ROM's outputs for a single input each. (**a**) Unit force at R_1 (**b**) Unit force at R_2 (**c**) Unit force at R_3 (**d**) Unit force at R_4 (**e**) Unit force at R_5 (**f**) Unit charge at the anode.

References

1. Chen, M.; Yu, H.; Guo, S.; Xu, R.; Shen, W. An electromagnetically-driven MEMS micromirror for laser projection. In Proceedings of the 10th IEEE International Conference on Nano/Micro Engineered and Molecular Systems, Xi'an, China, 7–11 April 2015; pp. 605–607. [CrossRef]
2. Fan, K.C.; Lin, W.L.; Chiang, L.H.; Chen, S.H.; Chung, T.T.; Yang, Y.J. A 2 × 2 Mechanical Optical Switch With a Thin MEMS Mirror. *J. Light. Technol.* **2009**, *27*, 1155–1161. [CrossRef]
3. Hung, A.C.L.; Lai, H.Y.H.; Lin, T.W.; Fu, S.G.; Lu, M.S.C. An electrostatically driven 2D micro-scanning mirror with capacitive sensing for projection display. *Sens. Actuators A Phys.* **2015**, *222*, 122–129. [CrossRef]

4. Cho, A.R.; Han, A.; Ju, S.; Jeong, H.; Park, J.H.; Kim, I.; Bu, J.U.; Ji, C.H. Electromagnetic biaxial microscanner with mechanical amplification at resonance. *Opt. Express* **2015**, *23*, 16792–16802. [CrossRef]
5. Yalcinkaya, A.D.; Urey, H.; Brown, D.; Montague, T.; Sprague, R. Two-Axis Electromagnetic Microscanner for High Resolution Displays. *J. Microelectromech. Syst.* **2006**, *15*, 786–794. [CrossRef]
6. Seo, Y.H.; Hwang, K.; Kim, H.; Jeong, K.H. Scanning MEMS Mirror for High Definition and High Frame Rate Lissajous Patterns. *Micromachines* **2019**, *10*, 67. [CrossRef]
7. Ju, S.; Jeong, H.; Park, J.H.; Bu, J.U.; Ji, C.H. Electromagnetic 2D Scanning Micromirror for High Definition Laser Projection Displays. *IEEE Photonics Technol. Lett.* **2018**, *30*, 2072–2075. [CrossRef]
8. Hwang, K.; Seo, Y.H.; Jeong, K.H. Microscanners for optical endomicroscopic applications. *Micro Nano Syst. Lett.* **2017**, *5*, 1. [CrossRef]
9. Seo, Y.H.; Hwang, K.; Jeong, K.H. 1.65 mm diameter forward-viewing confocal endomicroscopic catheter using a flip-chip bonded electrothermal MEMS fiber scanner. *Opt. Express* **2018**, *26*, 4780–4785. [CrossRef] [PubMed]
10. Hu, Q.; Pedersen, C.; Rodrigo, P.J. Eye-safe diode laser Doppler lidar with a MEMS beam-scanner. *Opt. Express* **2016**, *24*, 1934–1942. [CrossRef]
11. Kim, J.H.; Lee, S.W.; Jeong, H.S.; Lee, S.K.; Ji, C.H.; Park, J.H. Electromagnetically actuated 2-axis scanning micromirror with large aperture and tilting angle for lidar applications. In Proceedings of the 2015 Transducers—2015 18th International Conference on Solid-State Sensors, Actuators and Microsystems (TRANSDUCERS), Anchorage, AK, USA, 21–25 June 2015; pp. 839–842. [CrossRef]
12. Royo, S.; Ballesta-Garcia, M. An Overview of Lidar Imaging Systems for Autonomous Vehicles. *Appl. Sci.* **2019**, *9*, 4093. [CrossRef]
13. Wang, D.; Watkins, C.; Xie, H. MEMS Mirrors for LiDAR: A review. *Micromachines* **2020**, *11*, 456. [CrossRef]
14. Gu-Stoppel, S.; Giese, T.; Quenzer, H.J.; Hofmann, U.; Benecke, W. PZT-Actuated and -Sensed Resonant Micromirrors with Large Scan Angles Applying Mechanical Leverage Amplification for Biaxial Scanning. *Micromachines* **2017**, *8*, 215. [CrossRef] [PubMed]
15. Ou, C.H.; Lin, Y.C.; Keikoin, Y.; Ono, T.; Esashi, M.; Tsai, Y.C. Two-dimensional MEMS Fe-based metallic glass micromirror driven by an electromagnetic actuator. *Jpn. J. Appl. Phys.* **2019**, *58*, SDDL01. [CrossRef]
16. Park, Y.; Moon, S.; Lee, J.; Kim, K.; Lee, S.J.; Lee, J.H. Gimbal-Less Two-Axis Electromagnetic Microscanner with Twist Mechanism. *Micromachines* **2018**, *9*, 219. [CrossRef]
17. Jia, K.; Pal, S.; Xie, H. An Electrothermal Tip–Tilt–Piston Micromirror Based on Folded Dual S-Shaped Bimorphs. *J. Microelectromech. Syst.* **2009**, *18*, 1004–1015. [CrossRef]
18. Lara-Castro, M.; Herrera-Amaya, A.; Escarola-Rosas, M.A.; Vázquez-Toledo, M.; López-Huerta, F.; Aguilera-Cortés, L.A.; Herrera-May, A.L. Design and Modeling of Polysilicon Electrothermal Actuators for a MEMS Mirror with Low Power Consumption. *Micromachines* **2017**, *8*, 203. [CrossRef]
19. Markweg, E.; Nguyen, T.T.; Weinberger, S.; Ament, C.; Hoffmann, M. Development of a Miniaturized Multisensory Positioning Device for Laser Dicing Technology. *Phys. Procedia* **2011**, *12*, 387–395. [CrossRef]
20. Bunge, F.; Leopold, S.; Bohm, S.; Hoffmann, M. Scanning micromirror for large, quasi-static 2D-deflections based on electrostatic driven rotation of a hemisphere. *Sens. Actuators A Phys.* **2016**, *243*, 159–166. [CrossRef]
21. DFG. Kick and Catch—Cooperative Microactuators for Freely Moving Platforms: SPP 2206: Cooperative Multilevel Multistable Micro Actuator Systems (KOMMMA), 2019. Available online: https://gepris.dfg.de/gepris/projekt/424616052 (accessed on 16 June 2021).
22. Olbrich, M.; Schütz, A.; Kanjilal, K.; Bechtold, T.; Wallrabe, U.; Ament, C. Co-Design and Control of a Magnetic Microactuator for Freely Moving Platforms. In Proceedings of the 1st International Electronic Conference on Actuator Technology: Materials, Devices and Applications, Online, 23–27 November 2020; p. 8494. [CrossRef]
23. Schütz, A.; Hu, S.; Rudnyi, E.B.; Bechtold, T. Electromagnetic System-Level Model of Novel Free Flight Microactuator. In Proceedings of the 2020 21st International Conference on Thermal, Mechanical and Multi-Physics Simulation and Experiments in Microelectronics and Microsystems (EuroSimE), Online, 6–27 July 2020; pp. 1–6. [CrossRef]
24. Schütz, A.; Olbrich, M.; Hu, S.; Ament, C.; Bechtold, T. Parametric system-level models for position-control of novel electromagnetic free flight microactuator. *Microelectron. Reliab.* **2021**, *119*, 114062. [CrossRef]
25. Antoulas, A.C. *Approximation of Large-Scale Dynamical Systems*; Advances in design and control; Society for Industrial and Applied Mathematics: Philadelphia, PA, USA, 2005. [CrossRef]
26. Rudnyi, E.B. MOR for ANSYS. In *System-Level Modeling of MEMS*; Bechtold, T., Schrag, G., Feng, L., Eds.; Advanced Micro and Nanosystems; Wiley-VCH-Verl.: Weinheim, Germany, 2013; pp. 425–438.
27. Balajewicz, M.; Amsallem, D.; Farhat, C. Projection-based model reduction for contact problems. *Int. J. Numer. Methods Eng.* **2016**, *106*, 644–663. [CrossRef]
28. Nasdala, L. (Ed.) Kontakt. In *FEM-Formelsammlung Statik und Dynamik*; Springer Fachmedien Wiesbaden: Wiesbaden, Germany, 2015; pp. 227–244. [CrossRef]
29. Rutzmoser, J. Model Order Reduction for Nonlinear Structural Dynamics: Simulation-Free Approaches. Ph.D. Thesis, Technische Universität München, Garching, Germany, 2018.
30. Carlberg, K.; Bou-Mosleh, C.; Farhat, C. Efficient non-linear model reduction via a least-squares Petrov-Galerkin projection and compressive tensor approximations. *Int. J. Numer. Methods Eng.* **2011**, *86*, 155–181. [CrossRef]

31. Goury, O.; Duriez, C. Fast, Generic, and Reliable Control and Simulation of Soft Robots Using Model Order Reduction. *IEEE Trans. Robot.* **2018**, *34*, 1565–1576. [CrossRef]
32. Fauque, J.; Ramière, I.; Ryckelynck, D. Hybrid hyper-reduced modeling for contact mechanics problems. *Int. J. Numer. Methods Eng.* **2018**, *115*, 117–139. [CrossRef]
33. Tiso, P.; Rixen, D.J. Reduction methods for MEMS nonlinear dynamic analysis. In *Nonlinear Modeling and Applications*; Proulx, T., Ed.; Conference Proceedings of the Society for Experimental Mechanics Series; Springer: New York, NY, USA, 2011; Volume 2, pp. 53–65. [CrossRef]
34. Idelsohn, S.R.; Cardona, A. A reduction method for nonlinear structural dynamic analysis. *Comput. Methods Appl. Mech. Eng.* **1985**, *49*, 253–279. [CrossRef]
35. Manvelyan, D.; Simeon, B.; Wever, U. An Efficient Model Order Reduction Scheme for Dynamic Contact in Linear Elasticity. *arXiv* **2021**, arXiv:2102.03653.
36. Chaturantabut, S.; Sorensen, D.C. Nonlinear Model Reduction via Discrete Empirical Interpolation. *SIAM J. Sci. Comput.* **2010**, *32*, 2737–2764. [CrossRef]
37. Farhat, C.; Avery, P.; Chapman, T.; Cortial, J. Dimensional reduction of nonlinear finite element dynamic models with finite rotations and energy-based mesh sampling and weighting for computational efficiency. *Int. J. Numer. Methods Eng.* **2014**, *98*, 625–662. [CrossRef]
38. Farhat, C.; Chapman, T.; Avery, P. Structure-preserving, stability, and accuracy properties of the energy-conserving sampling and weighting method for the hyper reduction of nonlinear finite element dynamic models. *Int. J. Numer. Methods Eng.* **2015**, *102*, 1077–1110. [CrossRef]
39. Chapman, T. Nonlinear Model Order Reduction for Structural Systems with Contact. Ph.D. Thesis, Stanford University, Stanford, CA, USA, 2019.
40. Del Tin, L. Reduced-Order Modelling, Circuit-Level Design and SOI Fabrication of Microelectromechanical Resonators. Ph.D. Thesis, Università di Bologna, Bologna, Italy, 2007. [CrossRef]
41. Freund, R.W. Krylov-subspace methods for reduced-order modeling in circuit simulation. *J. Comput. Appl. Math.* **2000**, *123*, 395–421. [CrossRef]
42. Thorlabs, Inc. Piezo Actuators, Brochure. Available online: https://www.thorlabs.com/images/Brochures/Thorlabs_Piezo_Brochure.pdf (accessed on 16 June 2021).
43. Nasdala, L. (Ed.) *FEM-Formelsammlung Statik und Dynamik*; Springer Fachmedien Wiesbaden: Wiesbaden, Germany, 2015. [CrossRef]
44. Bathe, K.J. *Finite Element Procedures*, 2nd ed.; Prentice-Hall: Englewood Cliffs, NJ, USA, 2014.
45. Thorlabs, Inc. PA3JEAW-SpecSheet: Piezoelectric Chip, 100 V, 2.2 μm Displacement, $3.0 \times 3.0 \times 2.0$ mm, Pre-Attached Wires Available online: https://www.thorlabs.com/thorproduct.cfm?partnumber=PA3JEAW (accessed on 16 June 2021).
46. Bai, Z. Krylov subspace techniques for reduced-order modeling of large-scale dynamical systems. *Appl. Numer. Math.* **2002**, *43*, 9–44. [CrossRef]
47. Hu, S.; Yuan, C.; Bechtold, T. Quasi-Schur Transformation for the Stable Compact Modeling of Piezoelectric Energy Harvester Devices. In Proceedings of the 12th International Conference on Scientific Computing in Electrical Engineering, Taormina, Italy, 23–27 September 2018; pp. 267–276. [CrossRef]
48. Yuan, C.; Hu, S.; Castagnotto, A.; Lohmann, B.; Bechtold, T. Implicit Schur Complement for Model Order Reduction of Second Order Piezoelectric Energy Harvester Model. In Proceedings of the 9th Vienna International Conference on Mathematical Modelling (MATHMOD2018), Vienna, Austria, 21–23 February 2018. [CrossRef]
49. Hu, S.; Yuan, C.; Castagnotto, A.; Lohmann, B.; Bouhedma, S.; Hohlfeld, D.; Bechtold, T. Stable reduced order modeling of piezoelectric energy harvesting modules using implicit Schur complement. *Microelectron. Reliab.* **2018**, *85*, 148–155. [CrossRef]
50. Ansys®. Academic Research Electromagnetics Suite, Release 2020 R2, Canonsburg, PA, USA, 2020. Available online. https://www.ansys.com/products/release-highlights (accessed on 20 October 2021).

MDPI

St. Alban-Anlage 66

4052 Basel

Switzerland

Tel. +41 61 683 77 34

Fax +41 61 302 89 18

www.mdpi.com

Actuators Editorial Office

E-mail: actuators@mdpi.com

www.mdpi.com/journal/actuators